Jaqueline Maria Silva

Vitamin C supplementation in Brazilian native fish feed

Jaqueline Maria Silva

Vitamin C supplementation in Brazilian native fish feed

An antioxidant ingredient in fish feed

ScienciaScripts

Imprint

Cover image: www.ingimage.com

This book is a translation from the original published under ISBN 978-3-330-77228-1.

Publisher:
Sciencia Scripts
is a trademark of
Dodo Books Indian Ocean Ltd. and OmniScriptum S.R.L publishing group

120 High Road, East Finchley, London, N2 9ED, United Kingdom
Str. Armeneasca 28/1, office 1, Chisinau MD-2012, Republic of Moldova, Europe
Managing Directors: Ieva Konstantinova, Victoria Ursu
info@omniscriptum.com

Printed at: see last page
ISBN: 978-620-8-64127-6

SUMMARY

AUTHOR'S CURRICULUM VITAE

JAQUELINE MARIA DA SILVA - daughter of Heleno José da Silva (in memory) and Lenirà Maria da Conceiçâo, born on March 15, 1984, in the city of Arapiraca in the state of Alagoas. In 2003, she joined the Federal University of Alagoas (UFAL), where she graduated in Animal Science in January 2008. In February 2008, he began the Postgraduate Program in Animal Science, at Master's level, in the Department of Veterinary Medicine at the Federal Rural University of the Semi-Arid (UFERSA), in Mossoró-RN. He defended his dissertation in September 2009.

To God, for everything you are in my life.

To my parents, Heleno José da Silva (In memòria) and Lenira Maria da Conceiçao, for the opportunity, dedication, affection, support and love at every stage of my life. I love you eternally!

To my brother, José Edmilson da Silva, and my adopted sister, Cledjane dos Santos, for their affection and friendship.

To my cousins for their affection and support during this phase of my life.

To the Oliveira family, in the person of Mrs. Juligia, for their example of love, affection, companionship and life lessons.

ACKNOWLEDGMENTS

My special thanks go to Prof. Alex Martins Varela de Arruda for his guidance, patience and friendship.

To Prof·Silvia Maria Mendes Ahid, my thanks for the opportunity, the welcome, and the professional example of growth for those she works with.

To Prof. Édma Carvalho de Miranda, who has always been present in my life over the years. For her encouragement and belief in my ability, and for her technical and financial support in this research.

To Fernando Vianna Nobre for his friendship, support and encouragement. An admirable person with unparalleled professionalism, my special thanks, without whom none of this would be possible.

The Office of the Dean of Community Affairs - UFERSA, in the person of social worker Lucia and Dean Francisco Xavier de Oliveira Filho.

Thank you to Felipe, a research fellow at the laboratory, who was with us throughout this dissertation, for his dedication, friendship and good will. And to all the other colleagues in the laboratory who helped to complete this experiment.

To the Companhia de Desenvolvimento do Vale do Rio Sao Francisco e Paraiba (CODEVASF), for supplying the animals and physical space to set up the experiment.

To Mr. Francisco Rocha (CODEVASF-AL technical employee), for his technical support throughout the field experiment. Thank you for your friendship and longing.

To the master's students of the Postgraduate Program in Animal Sciences: Luciene Xavier, Ângela Gracindo, Nadja, Carlos Henrique, Felipe Barreto, Adza, Danielle, Nikolay, Thiberio de Souza, Francisco Vicente and Francisco David, for their help, affection and friendship.

My special thanks go to Luciene Xavier and Nadja, for their common affinities, companionship, complicity, affection, the friendship of sitting side by side, discussing work topics, in other words, united by a single thought "Producing Science". Thank you, girls, for your patience in listening to me talk about homesickness, about my family. This homesickness you helped me deal with. To Luciene, for giving me the opportunity to meet the wonderful lady who is Dona Lurdes, her sister and niece; to also meet Prof. Dr. Sakamoto, with his serene humility. I want you to know that you were part of this growth.

I thank Elba Carvalho for my spiritual and personal growth. Thanks again to Prof. Edma and to God for having met you. You are an angel in the life of anyone who meets you, and I can't thank you enough.

To Shirley (Shirha), for her support, friendship and companionship in this final moment of the dissertation, and to my new friends, Miguel (Faustao) with his joy; Paulo (O Véio) with his enthusiasm and determination; Kelly (Ligeirinha), with her care and affection for everyone, you are special.

To the girls from the UFERSA Academic Village: Juliana Oliveira, Kamila Queiroz, Natalia Santiago, Aline Menezes and Anne. Thank you for your affection and friendship.

To the Postgraduate Course in Animal Science and to all my colleagues, teachers and staff, for making this thesis possible.

Maria Luzia Nogueira, secretary of the UFERSA Post-Graduation Center, for helping me at various times with the documents required for the defense of this dissertation.

To CNPq / CAPES, for the master's scholarship and financial support.

To all those who contributed in some way to this work...

THANK YOU VERY MUCH.

EFFECT OF VITAMIN C SUPPLEMENTATION ON GROWTH RATIONS FOR THREE-SPOTTED PIAU *(LEPORINUS REINHARDIT)* - LUTKEN, 1874

SILVA, Jaqueline Maria da. **Effect of Vitamin C Supplementation in Growth Feeds for Piau Três Pintas (*Leporinus reinhardit*) - Lutken, 1874.** (**Dissertation**). Universidade Federal Rural do Semi-Arido (UFERSA), Mossoró-RN. 2009.

GENERAL ABSTRACT - The aim of this study was to evaluate the influence of vitamin C supplementation in the diet of three-spotted piau fish (*L. reinhardit*), in which performance and production yield were monitored, the nutritional chemical composition (carcass and fillet) was analyzed, as well as glycemic parameters. The field experiment lasted 90 days (fry to juvenile). A total of 1040 fry were used, housed in 16 rearing tanks with a capacity of 500 liters. The design used was a randomized block design with four treatments and four replications. The control feed was prepared to meet the vitamin C requirement according to commercial recommendations and the other feeds were prepared to contain extra dietary vitamin C supplementation levels (250, 500 and 750 mg ascorbic acid/kg feed) For the analysis of the nutritional chemical composition, 6 fish/experimental unit were used. All the diets were isoproteic and isocaloric (28% CP and 3100 kcal/kg of feed). There was no significant effect of the regression on production performance and glycemic parameters at the supplementary vitamin C levels. However, fillet yield showed a decreasing linear response to the effect of the treatments ($p<0.05$). With regard to the nutritional chemical composition, a significant response effect was observed for the vitamin C level of 750 mg and in other cases it was equal to the level of 60 mg. There was also an increasing linear response to the mineral matter content ($p<0.05$) and a decreasing trend for the crude energy content ($p<0.05$). Therefore, it can be concluded that under the conditions in which this work was carried out, the inclusion of vitamin C in the commercial level was sufficient to provide productive performance, benefits in terms of nutritional chemical composition and a reduction in handling and harvesting stress, as was the case for the feeds that received supplementation.

Keywords: Fish, Native, Vitamin C, Glucose, *Leporinus reinhardit.*

CHAPTER I - INITIAL CONSIDERATIONS

The Piau belongs to the family Anastomidae and the genus *Leporinus* sp. and is considered a little-known fish from a zootechnical point of view. However, some species of this genus, such as *L. elongatus*, *L. friderice*, *L. obtusidens*, *L. reinhardti* and *L. macrocephalus*, are being farmed in intensive systems in Brazil (RODRIGUES et al., 2006). Among the native fish that have potential for aquaculture in Brazil, the species of the genus *Leporinus* have the largest number of individuals in the river basins where they occur (GARAVELLO, 1979 apud CERICATO, 2005).

According to Duraes et al. (2001), the distribution of *Leporinus* species in southern Brazil is due to their opportunistic behavior, with *L. octofasciatus, L. elongatus, L. friderici* being found in greater numbers*, and L. amblyrhynchus, L. microphtalmus* and *L. paranensis* in smaller numbers. Several species of this genus are being cultivated, especially *L. macrocephalus*, which has the largest size within the genus and has been the most economically important species for fishing in the Pantanal of Mato Grosso, where it is known as "piavussu", but elsewhere called piauçu, piau, white piau (PERUCA et al., 2000).

In the wild, the piauçu has used a diverse and variable diet, depending on the season. However, it has been observed that piauçu in the post-larval period make good use of artificial food (NAVARRO et al., 2007; RODRIGUES et al., 2008). Hence the advantage of using this species in an intensive system. The species in this study, *L. reinhardti* (piau-très-pintas), is not widely cultivated because it has a migratory behavior and is abundant throughout the São Francisco River basin, including in its tributaries (ALVES & POMPEU, 2005).

Fish farming is characterized as an expanding economic activity in Brazil, especially in the Northeast, where the climate is conducive to continuous productivity throughout the year. In addition, the higher rate of population growth, with the consequent demand for food and the availability of areas suitable for fish farming, indicates the Northeast as a future development center for aquaculture (DIEGUES, 2006). However, like any other industrial process that aims to transform natural resources into consumer goods, aquaculture must be guided by the search for environmentally friendly technologies that meet the socio-economic and environmental development needs of low-income populations (FREITAS, 2006). The efficient and healthy development of fish in captivity depends on the supply of a diet capable of satisfying the basic needs for growth. In order to increase production, the first step is to determine the production-health ratio, which aims to determine the optimum animal productivity (NAVARRO et al., 2007).

1. 1 Vitamin C (Ascorbic Acid)

Vitamin C (ascorbic acid) is one of the nutrients present in feed, which acts as a strong reducing agent in various metabolic reactions. This vitamin is characterized by being water-soluble and thermo-instable (SILVA et al., 2007). Its importance for aquatic organisms is based on the form in which it is used, which is reduced or oxidized (L-ascorbic acid and dehydroascorbic acid), which represents the active form of greater retention in the body, and the less active form of simple elimination, respectively.

Fish and other animals need the enzyme L-gulonolactone oxidase to catalyze the last step in the transformation of glycuronic acid into ascorbic acid (Figure 1). This transformation occurs through D-glucose-1-phosphate activated by the stimulus caused by the binding of a nucleotide (uridine diphosphate - UDP), which is catalyzed by the enzyme glucose-1-phosphate uridyl transferase. After this process, UDP-glucose undergoes oxidation at carbon 6 (C-6) to form D-glucuronic acid, which is catalyzed by the enzyme UDP-glucose dehydrogenase. Through the hydrolysis process, D-glycuronic acid becomes D-glycuronate, acting as a precursor to ascorbic acid; D-glycuronate is reduced to L-gulonate, which is transformed into L-gulonolactone, then undergoes dehydrogenation by the flavoprotein L-gulonolactone oxidase to produce ascorbic acid (NELSON & COX, 2000).

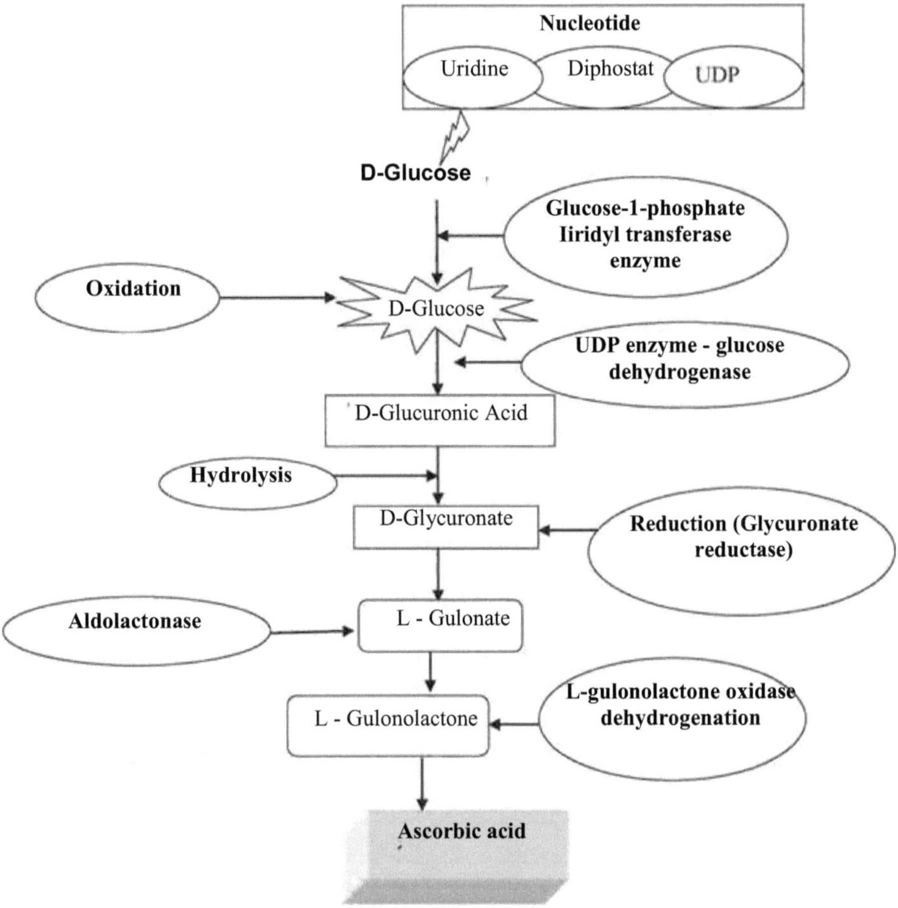

Figure 1. Ascorbic acid biosynthesis.

The preservation of ascorbate in tissues occurs through the reversal of dehydroascorbic into L-ascorbic acid, which can be found stored in the connective system of some fish (TACON, 1992 apud ALMEIDA, 2003). According to Fracalossi et al. (1998), initial studies postulated that vertebrates incapable of synthesizing vitamin C had an abundance of this nutrient in their natural diet and, consequently, did not need to preserve their biosynthetic ability. Thus, according to Cavichiolo et al. (2002a), as this vitamin is present in the water-soluble group, it is not stored in large quantities in the body, making it important to add it to the feed so that no signs of deficiency occur.

Bacconi (2003) describes that the greater the supplementation of this vitamin, the more efficient its absorption will be, due to the fact that the reaction in the body occurs via specific substrate-dependent transporters. According to Baldisserotto (2002) the cellular absorption of ascorbic acid occurs through simple and active facilitated diffusion, through the actions of the Na^+ / K^+ pump (concentration gradient in the enterocyte). Thus, Na^+ tends to enter and, as the transporter only works if there is a vitamin connected, it ends up carrying both into the cell. Ascorbic acid, in its reduced form, then diffuses from the inside of the enterocyte to the blood capillaries from the intestinal microvilli.

Vitamin C acts on the metabolism of fish, in terms of the structural (muscle and bone), immunological and reproductive systems, as a co-factor in the hydroxylation reactions of the amino acids proline and lysine, which are necessary for the formation of collagen present in connective tissues, in the healing and regeneration of injuries; it also acts in protein synthesis and is important for the formation of the bone matrix. This same hydroxylation reaction of amino acids is important for the formation and maintenance of adipose tissue (CHEN et al., 2004). This vitamin also acts in the stress response, in the biosynthesis of carnitine, which is involved in lipid metabolism and the peroxidation process, due to its antioxidant action, and as a vitamin E-sparing agent. The combined action of these vitamins acts directly on the metabolic and osmoregulatory balance, which enables organs and enzymes to function properly (HALVER, 2002).

In order to make better use of the actions of vitamin C in the animal organism, protected and stabilized forms of ascorbic acid have been studied (synthetic polymer, ethylcellulose, lipid/glyceride, ascorbic acid-2-sulphate, ascorbic acid-2-monophosphate, ascorbic acid-2-diphosphate, ascorbic acid-2-triphosphate, ascorbyl palmitate), aiming for greater stability and resistance to industrialization and storage processes (ROTTA, 2003). Among these forms, the most effective are ascorbic acid-2-sulphate and ascorbic acid-2-monophosphate (Figures 2 and 3).

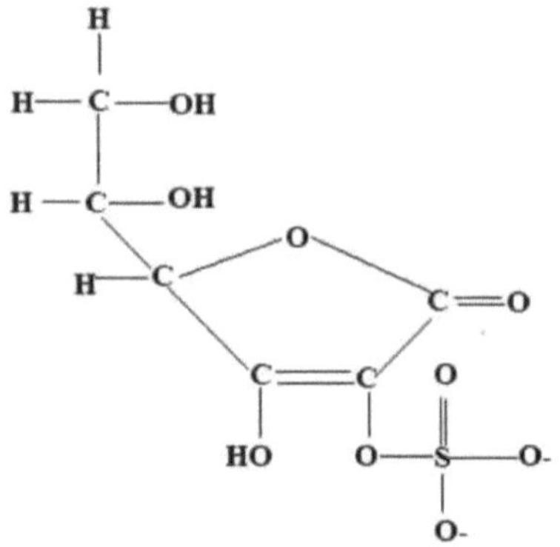

Figure 2: Stabilized form of ascorbic acid-2-sulphate.

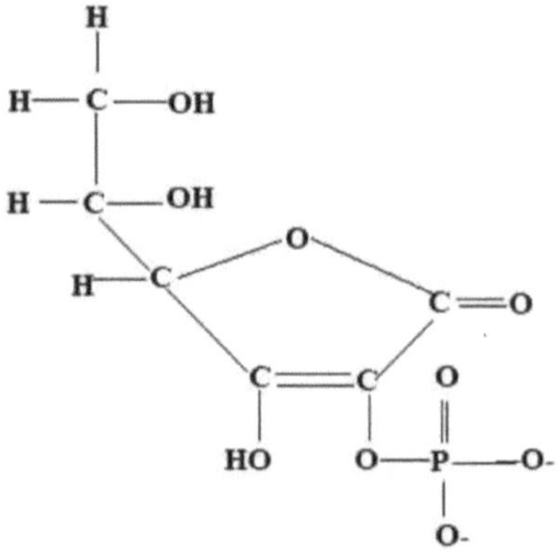

Figure 3: Stabilized form of ascorbic acid-2-monophosphate.

Some authors have studied ascorbyl palmitate because it retains its antioxidant characteristics and provides greater molecular stability, as well as acting in this protected way by enriching plankton and brine shrimp, which convert ascorbyl palmitate into ascorbic acid ROTTA (2003). According to Okamura et al. (2008) ascorbyl palmitate provides an increase in the total length and weight of gilthead seabream larvae, because like brine shrimp, fish larvae have shown the same conversion capacity.

On the other hand, the effect of the absence of ascorbic acid, as well as inadequate amounts in fish feed, can cause clinical signs of deficiency: reduced growth rate, morphological changes (muscle atrophy, gill deformities and opercular deformities), anemia, lethargy, darkening of the skin, reduced feed consumption, reduced immune resistance with susceptibility to infections (ectoparasites, bacteria and fungi), accumulation of fat in the liver, difficult healing (loss of scales with consequent damage to the skin), hemorrhages, anorexia and low resistance to stress (TACON, 1992; NRC, 1993).

Wang et al. (2002) observed signs of deficiency such as anorexia, scoliosis, hemorrhage, lower weight gain and lower protein efficiency in *Paralichthys olivaceus* fed a diet without vitamin C supplementation. Corroborating this, Chagas & Val (2003), working with tambaqui and vitamin C, observed that in the treatment without vitamin C the animals showed a reduction in growth. Falcon et al. (2007) also showed that the absence of vitamin C in Nile Tilapia diets impaired collagen synthesis, leading to irregular formation of the vertebrae, suggesting the onset of lordosis and scoliosis.

However, the supplementation of ascorbic acid in the diet improved the weight gain, feed conversion and survival of tambaquis after ten weeks of feeding (CHAGAS & VAL, 2003). Miranda et al. (2003) also found positive results for weight gain and survival of pacu larvae fed diets supplemented with vitamin C after 15 days of experimentation. Previously, Cavichiollo et al. (2002a) suggested adding vitamin C to diets for Nile tilapia, as it had a positive influence on survival rates.

Some authors have tried to determine the level of vitamin C requirements for certain fish species. Castagnolli et al. (1994) studied the vitamin C requirement for pacu (*Piaractus mesopotamicus*, Holmber, 1887) and found values of 50 mg / kg of feed. Tacon (1991) states that the range of recommended levels for the vast majority of fish species reared in the world varies from 10 to 1250 mg / kg. The literature shows that a level of 25 mg was sufficient to prevent a reduction in growth and the classic signs of vitamin deficiency in acarâ-açu (FRACALOSSI et al., 1998).

Mello et al. (1999) evaluated the response to hypoxia-induced stress in piauçu fry, fed diets supplemented with vitamin C, and reported that the intensity of the stress factor considered (low levels of dissolved oxygen in the water) was insufficient to cause fish mortality. According to Kubitza (1998), this species can withstand low levels of dissolved oxygen for several hours and is therefore resistant to transportation.

One of the major problems faced by fish farming is the spread of pathogens, which can occur through stress caused by transportation or naturally, through animals with deficient immune systems. According to Falcon (2007), the effect of vitamin C on the immune system is observed in the cells that carry out phagocytosis (leukocytes and macrophages) in which they produce substances to destroy the phagocytosed pathogens. In this way, vitamin C supplementation in fish feed could be used as an immunostimulant. As for natural infestation in tilapias, after 57 days of feeding increasing levels of vitamin C. Cavichiollo et al. (2002b) found greater resistance the higher the level of dietary supplementation. In the same vein, Wahli et al., (1998) reported a positive effect on the survival rate of rainbow trout fry subjected to ichio challenge stress and fed high levels of vitamin C. Borba et al., (2007) demonstrated that jundià fry supplemented with vitamin C, after 8 days of infestation with *Ichthyophthirius multifiliis*, showed 100% survival.

Vitamin C positively influenced the response to stress caused by the action of ectoparasites (*Trichodina*) in Nile Tilapia, reducing their percentage of occurrence (CAVICHIOLO et al., 2002a). Similar to what was observed in this study, Martins (1998) quantified the parasites in pacus, fed various levels of vitamin C, and found that the fish had the lowest numbers of parasites/fish. Martins et al. (2008) demonstrated in their study that vitamin supplementation enables greater migration of defense cells to the infected region. These results were in line with Sobhana et al., (2002) who observed greater infiltration of cells in *Cirrhinus mrigala*, supplemented with vitamin C in the infected region.

Fish don't always get sick and die when they are challenged by stressors. Normally, they adapt to the stress for a finite period of time and during this period, the fish may appear normal, but they are absorbing reserves and redirecting energy according to the extra demands imposed (BARTON & IWANA, 1991). In this condition, fish stop growing and reproducing and try to return to homeostasis. In these cases, there is a need to redistribute energy, and the animals must be well nourished to provide energy and other nutrients (vitamins) for the defense system to function properly.

If the stress is severe, prolonged and, above all, cumulative, the organism's adaptation may fail. This condition, according to Francis-Floyd (2002), becomes abnormal and is characterized by the gradual degeneration of the fish's ability to maintain its normal physiological condition. According to Martins et al. (2004), stress conditioned by the injection of carrageenan and lipopolysaccharide (LPS) into tilapia resulted in an increase in the percentage of neutrophils and a decrease in lymphocytes, when these animals were subjected to constant stress. This author reinforces that the response to stress can be considered species-specific, as this species showed a different response to the same type of stress when compared to *Piaractus mesopotamicus* and the tambaqui hybrid previously evaluated. According to the National Research Council (1993), these phagocytic cells produce reactive substances (free radicals) as a result of microbial action and the inflammatory response, which are toxic. However, vitamin C prevents damage to cells and surrounding tissues by activating the immune system, an important chemical mediator which can influence the inflammatory response.

The increase in metabolism during stress determines an increase in the requirement for vitamin C, in agreement with Montero et al, (1999) observed this increase in juvenile Gilthead sea bream (Sparus aurata) subjected to stocking density stress; as well as Falcon (2007) with Nile Tilapia subjected to cold stress; and by Borba et al., (2007) testing susceptibility to *Ichthyophthirius multifiliis* in jundià (*Rhamdia quelen*) fry.

The data presented makes the relationship between vitamin C and stress in fish very clear, revealing that levels above nutritional requirements have beneficial effects on resistance to disease. In research with Nile Tilapia larvae of Thai origin, Leonardo (1999) demonstrated a reduction in the occurrence of the ectoparasite *Trichodina* sp. from 72.5% in the treatment without vitamin C, to 52.5% in the treatment with 1000 mg of vitamin C. This highlights the importance of determining the vitamin C requirement for fish, given the growing diversity of intensive production systems. According to Al et al. (2004), vitamin requirements depend on the species, age, metabolic functions, feeding habits and farming system adopted.

Stressors such as: ectoparasites, temperature variations, hypoxia, transport, exposure time, method of exposure to the pathogen and handling are directly related to the mobilization of endogenous energy by the fish, i.e. increases in blood glucose levels (URBINATI et al., 2004). The purpose of glucose mobilization is to provide extra energy for the animal to resist during the period of disturbance. Ortuno et al., (2003) found that vitamin C supplementation in feed contained the rise in blood glucose in fish under stressful conditions.

Duccini et al., (2004) reported that changes in glucose levels are related to the level of activity of the species and the environment in which they are being raised. Cavero et al., (2004) used plasma glucose as an indirect indicator of stress in fish and pointed out that the levels increased when routine practices such as biometry and siphoning were carried out, reducing as the fish adapted to the experimental conditions. These results show that glucose can be used as an indicator of stress in fish under adverse conditions. However, Martins et al. (2008), looking for differences in the hematological picture and the acute inflammatory response induced in Nile tilapia, observed glucose levels and found no difference between treatments, which is possibly due to the animals' adaptation to the experimental conditions.

According to Baldisserotto (2002), fish adapt through a physiological adjustment or reorganization of cellular metabolism, with changes occurring in the synthesis of enzymes, in the concentration of substrate and/or metabolic pathway products and in the modulators of enzyme-substrate reactions, leading to an increase in glycogen reserves and a decrease in triacylglycerols in the fish's liver.

The alterations resulting from secondary responses to stress are marked by hydromineral imbalances, impaired immune functions and the mobilization of energy-rich substrates. These factors cause an immediate rise in glucose in response to the stressor, which can be caused by the stimulation of catecholamines, especially epinephrine, which stimulates glycogenolysis, i.e. the transformation of glycogen into glucose (BRANDÂO et al., 2006).

This transformation occurs through the mobilization of glycogen into glucose 1-phosphate, by means of a phosphorolytic reaction, which is catalyzed by glycogen phosphorylase. These enzymes catalyze the attack by inorganic phosphorus on the glycosidic bond (α 1 $\rightarrow$ 4), which joins the last two glucose residues at the non-reducing end, generating glucose 1-phosphate and a reduced polymer of one glucose unit. Phosphorylase preserves part of the energy of the glycosidic bond in glucose 1-phosphate, a phosphoric ester. Glycogen phosphorylase acts repeatedly until it approaches a branching point (α 1 $\rightarrow$ 6), where its action ceases. Glucose 1-phosphate, the end product of the glycogen phosphorylase reactions, is converted into glucose 6-phosphate by phosphoglucomutase, which catalyzes the reversible reaction: Glucose 1-phosphate θ Glucose 6-phosphate, which can enter the glycolytic pathway (LEHNINGER, 1995).

Glycogen is one of the many ways in which fish store the energy they consume as food. This energy source is found in large quantities in the liver and muscle tissues of fish (CYRINO et al., 2000). As a result of intense activity causing stress, the circulatory system of most small vertebrates, such as fish and other organisms, can transport oxygen to the muscles quickly enough to avoid the anaerobic consumption of muscle glycogen. However, when this activity takes place, the glycogen stores in the liver of the fish are consumed quickly and within a few minutes. Hepatic glycogen catabolism shows that this form of organic reserve is much more important as a supply of metabolic intermediates for biosynthetic processes than as a source of energy (CHRISTIANSEN & KLUNGSOYR, 1987). Glycogen is widely used in biochemical adaptations in various environmental stress situations.

In the physiological metabolism of fish, there is a metabolic regulator of glucose, insulin, which plays a secondary role in growth and cell differentiation in fish, although its exact functions in fish are still under investigation. According to studies by Plana et al. (2000) and Oliveira & Cyrino (2009), insulin has the same anabolic activity in fish as in mammals, reducing blood glucose levels, stimulating the incorporation of glycine into muscle and protein into the liver, and stimulating the mobilization of lipids and glycogen from muscle to liver. Sundby (1993) apud Bombardeli et al., (2003) describes the weight gain caused by the action of insulin on glucose in various species of fish, such as rainbow trout, Atlantic salmon and goldfish. Insulin causes a decrease in gluconeogenesis from amino acids and this saving of amino acids goes to the muscles for protein synthesis and deposition.

1. 2 Productive yield

Baldisserotto (2002) explains that the metabolic rate of fish as a function of a unit of weight (g or kg) is higher in small fish, and demonstrates that there is a close relationship between growth rate and energy expenditure (above the maintenance level).

Some authors report that the growth of the carcass in juvenile fish is greater than the growth of the other parts of the body, which tend to increase in weight. They also report that the relative size of tissues and organs, under adequate nutrition, is dependent only on the size of the fish and its life cycle (SHEARER, 1994 apud CYRINO et al., 2000). However, there are reports that fish fed different diets show different sizes and weights of the liver, in the most diverse species studied (CYRINO et al., 2000).

Smaller fish have better feed conversion, higher protein gain and lower fat gain when compared to larger animals (LOVELL, 1998 apud FALCON et al., 2007). According to the same author, the deposition of 1 g of fat represents a gain of 1 g, while the deposition of 1 g of protein represents a gain of 4 g. This is due to the fact that muscle is approximately 75% water, whereas fat is not.

Weight category and species influence muscle yield, as well as the definition of a cutting line for fillet removal (MACEDO - VIÉGAS et al., 1997 and SOUZA & MARANHAO, 2001). Others have shown that fish yield depends on anatomical structure, i.e. fish with large heads in relation to their muscles have lower fillet yields compared to those with small heads (EYO, 1993 and SOUZA & MARENGONI, 1998).

Carneiro et al. (2004), studying the processing of jundià (*Rhamdia quelen*), stated that females grew faster at weights above 500g and consequently the fillet yield was higher with fish of higher average weight.

Productive yield and processing studies show that several factors influence the yield after slaughter, such as: sex, size, anatomical shape of the body, size of the head and weight of viscera, skin and fins (CONTRERAS - GUZMAN, 1994 and MACEDO VIÉGAS et al., 1997). Gomiero et al. (2003) evaluated different types of matrinxâ head cut and observed carcass yields of over 50%, while fillet yields did not vary.

With regard to the types of nutrients, processing and storage used in the production and industrialization of fish, some factors can influence both the production yield and the chemical composition of the meat. According to Minozzo et al., (2002) the type of freezing adopted can influence the chemical composition of Nile tilapia fillets, and the same authors observed that glazing increases the nutritional content of fillets. Neto et al. (2006) tested different protein sources in *Leporinus obtusidens* and found that the carcass yield in all treatments was over 90%.

According to Reidel et al. (2004), who studied the influence of sex on the chemical composition of piavuçu and curimbatâ fillets, they observed that there was no relationship between the chemical composition of fillets and the different sexes of piavuçu and curimbatâ. Defining a feed that allows fish not only to grow quickly but also to adapt better has been the aim of several studies.

Captive breeding of native fish species is important and necessary due to the depletion of natural stocks and their vast consumer market. In addition, studies on the nutrition and feeding of native species are incipient, especially regarding the nutritional requirements of these animals; necessary tools to support the formulation of complete diets and consolidate the intensive and viable farming of native species.

2 REFERENCES

ALMEIDA, G. S. C. **Dietary supplementation of vitamin C, development and health of pacu (*Piaractus mesopotamicus*- Holmberg, 1887)**. 2003. p. 60. Dissertaçâo (Mestrado) - Escola Superior de Agricultura Luz de Queiroz. Piracicaba, SP. 2003.

AI, Q.; MAI, K.; ZHANG, C.; XU, W.; DUAN, Q.; TAN, B.; LIUFU, Z. Effects of dietary vitamin C on growth and immune response of Japanese sea bass, (*Lateolabrax japonicus*), **Aquaculture**, v. 242, p. 489-500. 2004.

ALVES, C. B. M. & POMPEU, P. S. Historical changes in the rio das velhas fish faunabrazil. IN: J. N. RINNE, R. M.; HUGHES, R. M. and CALAMUSSO, B. Editors. Historical changes in large river fish assemblages of the Americas. **American fisheries society**, Symposium 45, Bethesda, Maryland. p. 587-602. 2005.

BACCONI, D. F. **Nutritional requirement of vitamin A for Nile Tilapia (*Oreochromis niloticus*) fry**. 2003. Dissertation (Master's Degree) - Luiz de Queiroz College of Agriculture, Piracicaba, SP. 2003.

BALDISSEROTTO, B. **Fisiologia de peixes aplicada à piscicultura**. Santa Maria: UFSM, 211p., 2002.

BARTON, B. A.; IWANA, G. K.; Physiological changes in fish from stress in aquaculture with emphasis on the responses and effects of corticosteroids. **Annual Review of Fish Diseases (Journal Seek)**. v. 1, p. 3-26, 1991.

BOMBARDELLI, R. A.; MEURER, F.; SYPERRECK, M. A. Protein metabolism in fish. **Arquivos Ciência Vet. Zool.** Unipar, v. 7, n. 1, p. 69-79, Jan/Jun. 2003.

BORBA, M. R.; FRACALOSSI, D. M.; FREITAS, F. A. Efeito da suplementação de vitamina C na dieta sobre a susceptibilidade de alevinos de Jundià, Rhandia Quelen, ao (*Ichthyophthirius multifiliis*), **Acta Scienci animal scienci**, Maringà, v. 29, n. 1, p. 93 - 99, 2007.

BRANDÂO, F. R.; GOMES, L. C.; CHAGAS, E. C. Stress responses in pirarucu (*Arapaima gigas)* during routine fish farming practices. **Acta Amazônica**, v. 36, n. 3, p. 349-356, Aug. 2006.

CARNEIRO, P. C. F.; MIKOS, J. D.; BENDHACK, F.; IGNASIO, S. A. Processing of jundià (*Rhamdia quelen)* carcass yield. **Revista acadêmica**: Ciências Agràrias e Ambientais, Curitiba, v.2, n. 3, p. 11-17, Jul/Sep. 2004.

CASTAGNOLLI, N.; MARTINS, µ. L.; KRONKA, S. N. Effects of vitamin C supplementation on pacu (*Piaractus mesopotamicus* - HOLMBERG, 1887) nutrition. IN: BRAZILIAN AQUACULTURE SYMPOSIUM, 1992, Piracicaba. **ANAIS...**Piracicaba: FEALQ. p.239-251, 1994.

CAVERO, B. A. S.; PEREIRA-FILHO, M.; BORDINHON, A.M.; FONSECA, F. A. L.; ITUASSÙ, D. R.; ROUBACH, R.; ONO, E. A. Tolerance of juvenile pirarucu to increased ammonia concentration in a confined environment. **Pesquisa Agropecuària**, Brasilia, v.39, n.5, p. 513-516, Maio. 2004.

CAVICHIOLO, F.; VARGAS, L.; RIBEIRO, R. P.; MOREIRA, H. L. M.; LEONARDO, J. M.; Levels of vitamin C supplementation in feed on the occurrence of ectoparasites, survival and biomass in fry of Nile tilapia (*Oreochromis niloticus*). **Acta Scientiarum**, Maringà, v. 24, n. 4, p. 957 961, July. 2002a

CAVICHIOLO, F.; VARGAS, L.; RIBEIRO, R. P.; MOREIRA, H. L. M.; LOURES, B. R. R.; MAEHANA, K.; POVH, J. A.; LEONARDO, J. M. L. O. Effect of dietary vitamin C and vitamin E supplementation on ectoparasite occurrence, performance and survival in Nile Tilapia (*Oreochromis niloticus*) larvae during sex reversal. **Acta Scientiarum.** Maringà, v. 24, n. 4, p. 943-948, March. 2002b.

CERICATO, L. **Substitution of live food for artificial food and morphology of the digestive system of piau larvae, *Leporinus macrocephalus*.** 2005. Master's Degree. Paulista State University - UNESP Aquaculture Center. Jaboticabal - SP, 2005.

CONTRERAS-GUZMAN, E. S. **Biochemistry of fish and fish products**. Jaboticabal. FUNEP, 409p. 1994.

CYRINO, J. E. P.; PORTZ, L.; MARTINO, R. C.; Protein and energy retention in juvenile Black Bass (*Micropterus salmoides*). **Scientia Agricola**, v. 57, n. 4, p. 609616, Oct/Dec. 2000.

CHAGAS, E. O.; VAL, A. L. Effect of vitamin C on weight gain and hematological parameters of tambaqui. **Pesquisa Agropecuària Brasileira**, Brasilia, v. 38, n. 3, p. 397- 402. Jul/Aug. 2003.

CHEN, R.; LOCHMANN, R.; GOODWIN, A. et al. Effects of dietary vitamins C and E on alternative complement activity, hematology, tissue composition, vitamin concentrations and response to heat stress in juvenile golden shiner (*Notemigonus crysoleucas*). **Aquaculture**, v.242, p.553-569, 2004.

CHRISTIANSEN, D. C.; KLUNGSOYR, L. Metabolic utilization of nutrients and the effects of insulin in fish. **Compedium biochemical physiologic**, v.88, p. 701-711, 1987.

DIEGUES, A. C. Para uma aquicultura sustentâvel do Brasil. Nùcleo de apoio ά pesquisa sobre populaçôes humanas e àreas ùmidas brasileiras - USP, Art. N.3, 2006.

DUCCINI-SANTOS, C. P.; SANTOS-PERESTRELO, C.; AQUINO-SILVA, M. R.; GIRARDI, L.; FIORINI, M. P. Hematological study of Tilàpia nilòtica (*Oreochromis niloticus*) raised in net pens. IN: VIII LATIN AMERICAN ENCOUNTER OF SCIENTIFIC INITIATION AND IV LATIN AMERICAN ENCOUNTER OF POST-GRADUATION, 2004, **Anais...** University of Vale do Paraiba, 2004.

DURÂES, R.; POMPEU, P. S.; GODINHO, A. L. Feeding of four species of *Leporinus* (Characiformes, anostomidae) during the formation of a reservoir in southeastern Brazil. Iheringia, **Ser. Zool.**, Porto Alegre, v. 90: 183-191, 2001.

EYO, A. A. Carcass composition and filleting yield of tem fish species from kainji lake: Proceedings of the FAO expert consultation on fish technology in Africa. **FAO Fishers Report**, rome, n. 467, p. 173-175, 1993.

FALCON, D. R. **Lipid and vitamin C in dietary diets for Nile tilapia (*Oreochromis niloticus*).** 2007. p. 144. Dissertaçâo (Mestrado). Faculty of Veterinary Medicine and

Zootechny, Universidade Estadual Paulista, Botucatu. 2007.

FALCON, D. R.; BARROS, M. M.; PEZZATO, L. E.; VALLE, J. B. Lipid and vitamin C in winter preparatory diets for Nile Tilapia (*Oreochromis niloticus*).

Revista Brasileira de Zootecnia, v.36, n. 5, p. 1462-1472, 2007.

FRACALOSSI, D. M. Nutritional diseases in fish. IN: SYMPOSIUM ON FISH MANAGEMENT AND NUTRITION, Piracicaba. **Proceedings...** Campinas: CBNA, p. 97 - 122, 1998.

FRACALOSSI, D. M.; ALLEN, M. E.; NICHOLS, D. K.; OFTEDAL, O. T. Oscars, *Astronotus ocellatus*, have a dietary requirement for vitamin C. **Journal of Nutrition**, Bethesda, v. 128, p. 1745-1751, 1998a.

FRANCIS-FLOYD, R.; REED, P. ***Ichthyophthirius multifi,*** **(white spot) infections in fish**. Diseases of Aquatic Organisms. University of florida. p. 233-235, July, 2002.

FREITAS, R. R. Aquaculture allied to tourism as a leisure alternative and source of income. **Caderno Virtual de Turismo**, v. 6, n. 4, p. 1677 - 6976, 2006.

GARAVELLO, J. C. Taxonomic revision of the genus *Leporinus* Spix, 1829 (Ostariophys, Anostomidae), **Doctoral thesis** - Institute of Biosciences, University of São Paulo, São Paulo. 1979.

GOMIERO, J. S. G.; RIBEIRO, P. A. P.; FERREIRA, M. W.; LOGATO, P. V. R. Carcass yield of matrinxâ fish (*Bricon cephalus*) in the different head cuts. **Ciência Agrotecnologica**, Lavras, v. 27, n. 1, p. 211-216. Jan/Feb 2003.

HALVER, J. E. Fish nutrition. School of aquatic and fishery sciences. Washington: University of Seattle, 3 th. 2002, p. 99 - 107.

KUBITZA, F.; CYRINO, J. E. P.; ONO, E. A. Commercial fish feed in Brazil: Current situation and prospects. **Panorama da Aquicultura**, v. 8, n. 50, p. 38-49, 1998.

LEHNINGER, A. L.; NELSON, D. L.; COX, µ. M. **Principles of biochemistry**. São Paulo: Savier Editora de livros médicos LTDA, 1995.

LEONARDO, J. M. L. **O. Effect of vitamin C on Nile Tilapia (*Oreochromis niloticus*) larvae of Thai origin.** p. 55. 1999. Master's degree dissertation. State University of Maringà, Paranà, 1999.

LOVELL, R.T. **Nutrition and feeding of fish**. 2.ed. Massachusetts: Academic Press, 267p. 1998.

MARTINS, M. L. Evaluation of the addition of ascorbic acid to the ration of cultivated *Piaractus mesopotamicus* (CHARACIDAE) on the infrapopulation *of Anacanthorus penilabiatus* (MONOGENEA). **Brazilian journal of medical and biological research,** v. 31, p. 655-658, 1998.

MARTINS, M. L.; NOMURA, D. T.; MYIAZAKI, D. M. Y.; PILARSKY, F.; RIBEIRO, K.; CASTRO, M. P.; CAMPOS, C. F. M. Physiological and haematological response of *oreochromis niloticus* (Osteichthyes : Cichlidae) exposed to single and consecutive stress of capture. **Acta Scientiarum Animal Sciences**, Maringà, v. 26, n. 4, p. 449456, 2004.

MARTINS, M. L.; MIYAZAKI, D. M. Y.; MORAES, F. R.; GHIRALDELLI, L.; ADAMANTE, W. B.; MOURINO, J. L.P. Feed supplemented with vitamins C and E influences the acute inflammatory response in Nile tilapia. **Ciência Rural**, Santa Maria, v.38, n. 1, p. 213-218, Jan-Feb, 2008.

MACEDO-VIÉGAS, E. M.; SOUZA, M. L. R. Pre-processing and conservation of fish produced in fish farms. IN: CYRINO, J. E. P. et al. (Ed.) **Tópicos especiais em piscicultura de aqua doce tropical intensiva**. Sao Paulo, p. 405-500. 1997.

MELLO, R. F.; MOURA, M. A. M.; VIEIRA, I.; CYRINO, J. E. P. Supplementing the diet of piauçu (*Leporinus obtusidens*) fry with vitamin C. **Scientia Agricola**, v. 56, n. 4, p. 1223 - 1231, Oct./Dec. 1999 (supplement).

MIRANDA, E. C.; PINTO, L. G. Q.; FURUYA, W. M.; PEZZATO, L. E.; BARROS, M. M.; PEZZATO, A. C. Weight gain and survival rate of pacu (*Piaractus mesopotamicus*) post-larvae fed diets containing different levels of vitamin C. **Acta Scientiarum Animal Sciences**, Maringà, v.25, n. 1, p. 31-36. 2003.

MINOZZO, M. G.; VAZ, S. K.; GUBIANI, E. A.; JOHANN, A. P.; LAMPERTI, P.M.; MASSAGO, H.; BOSCOLO, W. Chemical composition of tilapia fillets (*Oreochromis niloticus* L.) subjected to freezing with and without glazing or chilling. IN: XI ENCONTRO ANUAL DE INICIAÇÂO CIENTÎFICA, 2002, Maringà, **ANAIS...** Maringà - State University of Maringà, 2002.

MONTERO, D.; MARRERO, M.; IZQUIERDO, M. S.; ROBAINA, L.; VERGARA, J. M.; TORT, L. Effect of vitamin E and C dietary supplementation on some immune parameters of gilthead seabream (*Spaurus aurata*) juveniles subjected to crowding stress. **Aquaculture**, v. 171, p. 269-278, 1999.

NATIONAL RESEARCH COUNCIL (NRC). **Nutrient requirements of fishes Washington DC:** National academy of sciences, 1993.

NAVARRO, R. D.; LANNA, E. A. T.; DONZELE, J. L.; MATTA, S. L. P.; SOUZA, M. A.. Levels of digestible energy in the diet on the performance of piauçu (*Leporinus macrocephalus*) in the post-larval stage. **Acta Sci. Anim. Sci.** Maringà, v. 29, n. 1, p. 109114. 2007

NELSON, D. L.; COX, M. M. **Lehninger**: Principles of biochemistry. 3 ed. New York: Worth publishers, p. 1152, 2000.

NETO, J. R.; LAZARRI, R.; PEDRON, F. A.; VEIVERBERG, C. A.; BERGAMIN, G. T.; CORRÊIA, V.; FILIPETTO, J. E. S. Feeding of piava (*Leporinus obtusidens)* with different protein sources. **Ciência Rural**, Santa Maria, v. 36, n. 5, p. 1611-1616. Sept/Oct. 2006.

ORTUNO, J.; ESTEBAN, J.; MESEGUER, M. A. The effect of dietary intake of vitamin C and E on the stress response of gilthead seabream (*Sparus aurata L.*) **Fish Shellfish Immunol**, v.14, p. 145-156. 2003.

OLIVEIRA, A. M. B. M. S.; CYRINO, J. E. P.. Hormonal regulation of growth in fish. Available at: http://www.pisciculturapaulista.com.br/pdf/

regulation_hormonal_growth_fish.pdf. Accessed on: June 25, 2009.

OKAMURA, D.; ARAUJO, F. G.; LOGATO, P. V. R.; SILVEIRA, U. S.; MURGAS, L. D. S.; FREITAS, R. T. F. Ascorbyl palmitate and tocopherol acetate as metabolic antioxidants in gilthead seabream larvae. **Pesquisa Agropecuària Brasileira**, Brasilia, v.43, n.8, p. 1061-1068, Aug. 2008.

PLANA, J. V.; MENDEZ, E.; BANOS, N.; CAPILLA, E.; CASTILLO, J.; NAVARRO, I.; GUTIÉRREZ, J. Fish insulin, IGF-I and IGF-II receptors: Aphylogenetic approach. **Amin. Zool**. v. 40, p. 223-233, 2000.

PERUCA, A. P. S.; TEXEIRA, I.; BENNEMANNS, S. T. Frugivory in juveniles of three species of fish of the genus *Leporinus*, IN: XXIII BRAZILIAN CONGRESS OF ZOOLOGY. Cuiabà. **ANAIS...** Cuiabà. p. 381, 2000.

REIDEL, A.; OLIVEIRA, L. G., PIANA, P. A.; LEMAINSKI, D.; BOMBARDELLI, R. A.; BOSCOLO, W. R. Evaluation of yield and morphometric characteristics of male and female curimbatà *Prochilodus lineatus* (Valenciennes, 1836), and pivuçu *Leporinus macrocephalus* (Garavello & Britski, 1988). **Revista Varia Scientia**, Cascavel, v.4, n.8, p.71-78, 2004.

ROTTA, M. A. **Utilization of ascorbic acid (vitamin C) by fish**. Corumbà: Embrapa pantanal, 2003. p. 54, (Documentos / Embrapa pantanal, ISSN 1517-1973; 49), 2003.

RODRIGUES, S. S.; NAVARRO, R. D.; MENIN, E. Anatomical adaptations of the buccopharyngeal cavity of *Leporinus macrocephalus* Garavello and Britski, 1988 (Characiformes, Anostomidae) in relation to feeding habits. **Revista biotemas**, v. 19, n.1, p.51-58, 2006.

RODRIGUES, S. S.; NAVARRO, R. D.; MENIN, E. Digestive tract anatomy *of Leporinus macrocephalus* GARAVELLO & BRITSKI, 1988 (Characiformes, anostomidae) in relation to its feeding habit. **Journal Biosci**., Uberlândia, v. 24, n. 3, p. 86 - 95, July / Sept., 2008.

SILVA, L. S.; VICENZI, R.; PEIXOTO, C. R. M. Effect of malate ion on the oxidation process of ascorbic acid by molecular oxygen catalyzed by copper II ions. **Eclética quimica**, Sâo Paulo, v. 32, n. 2, p. 35-40, 2007.

SOBHANA, K. S. ET AL. Effect of dietary vitamin C on the inflammatory respone of mrigal, (*Cirrhinus mrigala* - HAMILTON), to freund's complete adjuvant. **Journal of fish diseases**, Stirling, v. 25, p. 179-184, 2002.

SOUZA, M. L. R.; MARANHAO, T. C. F. Carcass, fillet and filleting by-product yield of Nile tilapia (*Oreochromis niloticus* L.) as a function of body weight. **Alta Scientiarum.** Maringà, v. 23, n. 4, p. 897-901, 2001.

SOUZA, M. L. R.; MARENGONI, N. G. Processing yeld of nile tilàpia (*Oreochromis niloticus*) and channel catfish (*Ictalurus punctatus*). IN: PANAMERICAN CONGRESS OF VETERINARY SCIENCES, 16. 1998. Santa Cruz de La Sierra, Bolivia, Memorias **Summary...** Santa Cruz de La Sierra, 1998. p. 166.

SHEARER, K.D. Factors affecting the proximate composition of cultured fishes with emphasis on salmonids. **Aquaculture**, v.119, p.63-88, 1994.

SUNDBY, S. Dispersal and growth of pelagic juvenile arcto-norwegian cod (*Gadus morhua*) inferred from otolith microstructure and water temperature. International council for the exploration of the sea, **Journal Mar. Science.** v. 50, p. 261-270, 1993.

TACON, A. G. J. Vitamin nutrition in shrimp and fish. In: Aquaculture feed. PROCESSING AND NUTRITION WORKSHOP, Singapore, **Proceedings...** Singapore: American Soybean Association, p. 11-41, 1991.

TACON, A. G. J. **Nutritional fish pathology:** morphological signs of nutrient deficiency and toxicity in farmed fish. Rome: FAO, p. 75, 1992.

URBINATI, E. C.; ABREU, J. S.; CAMARGO, A. C. S.; LANDINES, M. A. Loading and transport stress in juvenile matrinxa (*Brycon cephalus*) at various densities. **Aquaculture**. v. 229, p. 389-400. 2004.

WAHLI, I.; VERLHAC, V.; GABAUDAN, J. ET AL. Influence of combined vitamin C and E on non-specific immunity and disease resistance of rainbow trout (*Oncorhynchus mykiss* - WALBAUM). **Journal of fish diseases,** v.21, p. 127-137, 1998.

WANG, X. et al. Effects of different dietary levels of L - ascorbyl -2-polyphosphate on growth and tissue vitamin C concentrations in juvenile liver flounder, *Paralichthys olivaceus* (TEMMINCK ET SCHLEGEL). **Revista Aquaculture,** v. 33, p. 261 - 267, 2002.

CHAPTER 2 - EFFECT OF VITAMIN C SUPPLEMENTATION ON THE GROWTH OF THE THREE-PAWED *GUINEA* FOWL (*LEPORINUS REINHARDTI*) - LUTKEN, 1874

ABSTRACT - To evaluate the dietary supplementation of vitamin C in growth diets for three-spotted piau fish (*Leporinus reinhardit*), on performance and production yield, body chemical composition (carcass and fillet) and analysis of glycemic parameters. A total of 1040 fry were used, distributed at a density of 130 fish/m^3 in experimental tanks for 90 days, in a randomized block design, with four treatments and four replications. The animals were fed isoprotein and isocaloric diets (28% CP and 3100 kcal/kg of feed). The control diet was designed to meet the vitamin C requirement according to commercial recommendations and the other diets were designed to contain extra dietary vitamin C supplementation levels (250, 500 and 750 mg ascorbic acid/kg feed). There was no significant effect through regression ($p>0.05$) for different levels of vitamin C in the diet on weight gain (WG), feed conversion (FC), specific growth rate (SGR), protein efficiency rate (PER), carcass yield (CR), somatic digestive index (SDI), survival (SBV), hepato-somatic index (HSI), glucose (GLI) and glycogen (GCO). Fillet yield (FR) showed a decreasing linear response ($p<0.05$) with an average value of 35.32%. Body chemical composition showed a significant linear effect for dry matter ($p<0.05$) with an average value of 98.11%, for mineral matter there was an increasing linear effect ($p<0.05$) with an average value of 19.60%, for crude energy there was a quadratic effect of vitamin C levels with ($p<0.05$) and the average EB value was 5486.71 kcal/kg. There was no significant effect on glycemic parameters in the regression analysis, but the blood glucose level showed a significant difference in the analysis of variance with an average value of 53.62 mg/dL. Higher levels of extra dietary supplementation show benefits in terms of performance, body composition and a reduction in production and/or harvesting stress.

Keywords: Ascorbic Acid, Nutrition, Fish, Health

2. 1 INTRODUCTION

Brazilian fish farming lacks technical information on the intensive cultivation of native species, making it difficult to develop production on a large commercial scale. However, recent research has provided relevant information on adaptability and performance, nutritional requirements and feeding habits (OKAMURA et al., 2008).

The three-spotted piau (*Leporinus reinhardti* - Lutken, 1874) is a migratory fish that inhabits the São Francisco River basin and its tributaries, but can also live and reproduce in dams

(ALVES & POMPEU, 2007). Currently, among the possibilities for zootechnical exploitation of native fish, the *Leporinus* genus has stood out due to its good adaptation to intensive farming and good meat quality (NAVARRO et al., 2007). These fish adapt to diets quite easily depending on the availability of food, allowing for different nutritional strategies in captivity, with the use of vitamin and mineral supplements (NETO et al., 2006).

Intensive production systems that generate good quality products require a correct balance of feed in order to be effective and produce with environmental responsibility, so the use of micronutrients, immunostimulants and various dietary chelated compounds are used to provide the best use of nutrients by the animal. In this context, vitamin C is considered an immunostimulant micronutrient, which acts in the most diverse metabolic reactions of the animal organism, as co-factors or specific substrates, aiming at growth, health and reproduction (BACCONI, 2003). Some fish do not synthesize vitamin C due to the lack of the enzyme L-gulonolactone oxidase for its synthesis from glucose (LEHININGER et al., 1995).

According to Fracalossi et al., (1998) initial studies postulated that animals incapable of synthesizing vitamin C had an abundance of this nutrient in their natural diet, and thus did not preserve their biosynthetic ability. According to Cavichiolo et al. (2002a), this water-soluble vitamin is not stored in large quantities in the body, so it is important to constantly add it to the feed to avoid signs of deficiency; the absence or excess of this vitamin in the body can cause reduced weight gain, scoliosis, lordosis, kyphosis, deformities in the head and operculum, erosion of the caudal fin, anemia, hemorrhages, limitations in protein and energy retention.

Vitamin C requirements vary according to species, growth stage, diet and environment, and smaller fish have a higher vitamin C requirement than larger fish, suggesting that the need for vitamin C decreases with age (AI et al., 2004).

The growth and body chemical composition of fish can be influenced by manipulating the different feeds and nutrients supplied to them. Several studies have described the body chemical composition of fish fed different diets, and there is little information on the effect of vitamin C supplementation on the composition of the protein and energy reserves of different tissues in fish (JOBLING et al., 1998).

Studies related to the nutrition of tropical fish involve knowledge of the nutritional requirements and the variables of the chemical composition of the fillet and carcass, since the industrialization of fish requires such knowledge due to the influence on the freezing and storage processes, as well as the effects on the organoleptic characteristics of the fish

(GONÇALVES et al., 2002).

In this way, poorly formulated feeds can cause serious environmental impacts and reduce production potential. Therefore, the aim of this study was to evaluate the effects of vitamin C supplementation in the growth diets of three-spotted piau fish (*Leporinus reinhardti)* on their performance and production yield, body chemical composition (fillet and carcass) and glycemic parameters.

2. 2 MATERIAL AND METHODS

The experiment was carried out at the Reference Center for Fisheries Resources and Aquaculture (CERAQUA) of the Companhia de Desenvolvimento dos Vales do Rio Sao Francisco e do Paraiba (CODEVASF), in Porto Real do Colégio - Alagoas, between January and April 2009, with a total experimental period of 90 days. A total of 1040 fry were used, with an average weight of 0.4392 ± 0.1549 g and an average length of 2.70± 0.20 cm. The fry were weighed and measured at the beginning and end of the experiment and distributed to the tanks at a density of 130 fish/m^3. The experimental units consisted of batches of fry placed in 16 polyethylene tanks, each with a total volume of 500L, supplied by a continuous water system. Water quality parameters such as dissolved oxygen and temperature were monitored once a week using a portable oximeter, and hydrogen potential (pH) using a portable pH meter.

The experiment was laid out in randomized blocks, with four treatments and four replications. The control feed was designed to meet commercial recommendations (Table 1) and the other feeds were designed to contain increasing levels of extra dietary vitamin C supplementation (250; 500 and 750 mg / kg feed). All the experimental rations were balanced to be isoproteic and isocaloric (28% CP and 3100 kcal / kg of feed), and extruded according to the recommendations of the NRC (1993), i.e. the rations contained the same ingredients and nutrients, differing only in the supplementation of ascorbic acid (Ascorbyl-2-monophosphate - 35% activity), and the vitamin premix used was free of vitamin C. The vitamin concentration of each treatment was calculated based on the vitamin availability of the product used.

The fry were fed *ad libitum* at three meals a day (8:00 am, 12:30 pm and 4:30 pm) for a period of 90 days, and consumption was monitored every 15 days using biometry. Feed was distributed based on weighing and observations of voluntary consumption (fish satiety behavior).

Table 1. Percentage and chemical-bromatological composition of experimental treatments

Ingredients (%)	Vitamin C levels mg/kg 60mg	250mg	500mg	750mg
Soybean meal	36,70	37,00	37,00	37,30
Cornmeal	28,57	27,53	26,40	25,00
Feather Flour	8,50	8,50	8,50	8,50
Wheat bran	13,50	13,50	13,50	13,50
Quirera Rice	3,00	3,00	3,00	3,00
L - Lysine	0,42	0,42	0,42	0,42
DL - Methionine	0,33	0,33	0,33	0,33
Vit C[1]	0,17	0,71	1,43	2,14
Soybean oil	1,85	2,05	2,47	2,85
Bi-Calcium Phosphate	3,70	3,70	3,70	3,70
Calcium	2,50	2,50	2,50	2,50
Common salt	0,50	0,50	0,50	0,50
Vitamin Premix[2]	0,06	0,06	0,06	0,06
Mineral premix[3]	0,18	0,18	0,18	0,18
BHT [4]	0,02	0,02	0,02	0,02
Total	100,00	100,00	100,00	100,00
Feed composition **(%):**				
Digestible energy (kcal/kg) [5]	3105,40	3101,97	3103,14	3103,78
Crude Protein	28,03	28,09	28,01	28,04
Crude fiber	4,11	4,11	4,09	4,08
Ethereal extract	4,90	5,06	5,42	5,74
Total calcium	2,05	2,05	2,05	2,05
Available phosphorus	0,60	0,60	0,60	0,60
Methionine[5]	0,70	0,70	0,70	0,70
Sulfur Amino Acids	1,32	1,32	1,31	1,31
Lysine[5]	1,70	1,70	1,70	1,71
Tryptophan[5]	0,33	0,33	0,33	0,33
Vit C (mg/kg)	60	250	500	750

[1]Vit. C = 35% active ingredient of ascorbium-2-monophosphate;

[2] Vitamin Premix: guarantee levels per kg of product: Vitamins: A = 1200,000 IU; D3= 200,000 IU; E = 12,000 mg; K3 = 2,400 mg; B1 = 4,800 mg; B2 = 4,800 mg; B6 = 4,000 mg; B12 = 4,800 mg; ac. Folic = 1,200 mg; Ca pantothenate = 12,000 mg; biotin = 48mg; choline = 65,000mg; niacin = 24,000mg;

[3] Mineral premix: iron = 10,000 mg; copper = 600 mg; manganese = 4,000 mg; zinc = 6,000 mg; iodine = 20 mg; cobalt = 2 mg and selenium = 20 mg;

[4] BHT = Butylated hydroxy toluene (antioxidant);

[5] Estimated values based on A.O.A.C. (1984).

During the experimental period, mortality was recorded and at the end of the experiment, all the animals in each experimental unit were counted and weighed. Before slaughter, one animal was taken from each experimental unit to assess blood glucose levels (mg / dL) using a digital glucometer (Accu-Check Active Glucose Monitor - Roche), in which a drop of blood was added to the strip for immediate reading.

For slaughter, the animals were subjected to heat shock in a container with ice and water (CARNEIRO et al., 2004). In order to obtain the productive performance variables, Weight Gain (WG g); Feed Conversion (FC g / g); Specific Growth Rate (SGR %); Protein Efficiency Rate (PER %) and Survival (SBV %), the following calculations were made:

$$GP = (Pf - Pi) \qquad TCE\ (\%dia^{-1}) = \frac{(\ln Pf - \ln Pi)}{(\Delta t)} x\ 100$$

$$TEP\ (\%) = \frac{GP}{CR\ x\ \%PB\ dieta} x\ 100 \qquad SBV\ (\%) = \frac{(N^{o}\ AN.\ Finais\ x\ 100)}{(N^{o}\ AN.\ Iniciais)}$$

Where *Pi* = Average initial weight; *Pf* = Average final weight; Δt = Experiment time in days; *CR* = Breed consumption; *PB%*= Crude protein; *N° AN.*= Number of animals.

To obtain the values for Somatic Digestive Index (SDI %); Hepatic-Somatic Index (HSI %); Fillet Yield (FR %) and Carcass Yield (CR %), 10% (six fish) were taken from each experimental unit. The carcass was then weighed, the fillet removed and the weight obtained, packed in plastic bags, duly labeled and kept frozen until the analysis was completed. The fillet was obtained from the dorsal muscles, lengthwise, along the entire length of the spine and ribs (CARNEIRO et al., 2004 and OMENA, 2008). The data was submitted, in accordance with SILVA & ANDERSON (1995) and COSTA et al., (2008) to the following formulas:

$$IDS\ (\%) = \frac{(Peso\ do\ trato\ digestorio)}{(Peso\ final)} x\ 100 \qquad HIS\ (\%) = \frac{(Peso\ do\ fígado)}{(Peso\ final)} x\ 100$$

$$RF\ (\%) = \frac{(Peso\ de\ 2\ filés)}{(Peso\ inteiro)} x\ 100 \qquad RC\ (\%) = \frac{(Peso\ eviscerado)}{(Peso\ inteiro)} x\ 100$$

The livers were then weighed and placed in an ependoff tube, duly labeled and immediately frozen for later analysis of liver glycogen using colorimetric analysis.

The fillet and carcass samples were ground in a micro-mill for total homogenization and analysis sequences. The analyses of the chemical composition (dry matter , mineral matter, crude protein, ethereal extract and crude energy) were carried out at the Applied Enzymology and Bromatological Analysis Laboratory of the Federal University of Alagoas, and followed the technical determinations proposed by SILVA & QUEIROZ (2002), while the determination of liver glycogen followed the method proposed by BIDINOTTO et al, (1997) and DUBOIS et al., (1960).

The data obtained was subjected to analysis of variance and linear and polynomial regression using SAS software® (Statistical Analysis System).

2. 3 Results and discussion

The water quality parameters used in the aquariums remained within the acceptable range for tropical fish, with average values in the morning for temperature, dissolved oxygen and pH of 28.0± 0.58 °C; 5.0± 0.81 mg/L and 7.5± 0.37, respectively, and in the afternoon, 28.4 ± 0.79 °C; 4.9 ± 0.50 mg/L; 7.5 ± 0.44, respectively. These values can be considered adequate and probably did not influence the variables analyzed in this study.

Table 2 shows the performance parameters, weight gain (WG), feed conversion (FC), specific growth rate (SGR), protein efficiency rate (PER) and survival (SSS), which showed no significant effects of the treatments in the regression analysis. The action of vitamin C on production performance is still controversial, especially in relation to weight gain, feed conversion and survival rate (BARROS et al., 2002; KASAY, 2007).

Table 2. Performance of three-spotted piau fish fed supplementary levels of vitamin C in their diets.

Performance parameters	Vitamin C levels mg/kg 60	250	500	750	Effect ■ linear	Quadratic effect
GP (g)[1] dp	35,92 (± 10,11)	33,37 (± 7,99)	25,57 (± 5,44)	38,77 (± 9,35)	ns	ns
CA (g/g)[2] dp	1,37 (± 0,39)	1,20 (± 0,22)	1,26 (± 0,31)	1,10 (+ 0,26)	ns	ns
TCE (%)[3] dp	4,81 (± 0,33)	4,77 (± 0,23)	4,48 (± 0,25)	4,90 (± 0,27)	ns	ns
TEP (%)[4] dp	3,17 (± 0,89)	3,17 (± 0,89)	3,11 (± 0,66)	3,88 (± 0,94)	ns	ns
SBV (%)[5] dp	65,00 (± 16,96)	73,85 (± 15,58)	90,38 (± 9,59)	73,46 (± 20,15)	ns	ns

(GP)[1]: Weight gain; (CA)[2]: Feed conversion; (TCE)[(3)]: Specific daily growth rate; (TEP)[(4)]: Protein efficiency rate; (SBV)[(5)]: Survival rate; sd (± standard deviation).

Table 2 shows that the 750 mg/kg level of vitamin C gave the best GP and AC values. Although there was no statistically significant effect from the regression analysis. According to Dabrowski et al. (1994), the greater the supplementation of this vitamin, the more efficiently it is absorbed and acts in the body as a co-factor for various reactions, including protein synthesis. This same behavior was observed by FUJIMOTO & CARNEIRO (2001) with guinea fowl, in which they explained that this fact may have occurred due to the high heterogeneity in the growth of this species, which resulted in high values for the coefficients of variation.

In experiments with the same genus of fish as in the present study, *Leporinus*, Mello et al. (1999) found no significant difference in the variables of productive performance, suggesting that the chemical form of vitamin C in the diet, the particle size of the feed and the quality of the water in the experiment may have favored the absence of effects in the treatments. In addition, the feeding behavior of these fish may have been influenced by management, with a pattern of late ingestion being observed after the feed was given, and the length of time the feed remained in the water may have favored a certain loss of vitamin C, in agreement with the inferences made by (TACON, 1991).

The feed conversion in this study showed no significant difference (Table 2), corroborating the results found in guinea fowl (*Pseudoplatystoma corruscans*) fed 0 to 250 mg of vitamin C / kg by FUJIMOTO & CARNEIRO (2001); and with Nile tilapia (*Oreochromis niloticus*) fed 0 to 115 mg of vitamin C / kg by BARROS et al, (2002); and in pacu fry (*Piaractus mesopotamicus*) fed diets supplemented with 0 to 900 mg of vitamin C / kg by ALMEIDA (2003).

In this study, the mean EER and TEP were not significantly affected by the treatments, but the values obtained can be considered high, ranging from 4.90 to 4.48 % per day and 3.88 to 3.11 %, respectively. According to Almeida (2003), values above 2.5 % per day represent high growth rates, corroborating Luna-figueroa (2003), who obtained even higher EER (4.19 to 6.29 % per day), reporting that fish in the early stages of growth can provide high average EER. Zuanon et al. (2006) obtained values of 2.31 to 2.61 % per day when working with flagtail fry. As for the values recorded for TEP, these were similar to the results obtained by Souza et al., (2004) and higher than those obtained with *Leporinus macrocephalus* by Soares et al., (2000), suggesting a good growth pattern obtained by *Leporinus reinhardti* in an intensive rearing system. According to the relevant literature, the results are inconsistent for vitamin C supplementation levels for fish native to Brazil, and do not yet allow us to establish nutritional recommendations that are different from dietary safety margins, such as for trairâo (MARTINS, 1994; KASAY, 2007) and pacu (ALMEIDA, 2003).

For the survival rate of the fish, it was found that all levels of vitamin C supplementation led to better SBV rates, with the 500 mg / kg level standing out with (90.38%), compared to the lowest supplemented level of 60 mg / kg, similar to that found by Mello et al., (1999), with higher survival at the 550 mg / kg level of vitamin C supplementation. Miranda et al. (2003) found higher survival when they supplemented the feed with the lowest level of vitamin C (250 mg / kg).

The non-significant differences in the survival of *L. reinhardit* are similar to the results found by other authors, such as Cavichiolo et al., (2002[a]), Chagas & Val (2003), Kasay (2007) who studied Nile tilapia, tambaqui and trairao, respectively.

Table 3. Carcass and fillet yield and physiological parameters of three-spotted piau fish fed supplementary levels of vitamin C in their diets.

Parameters	Vitamin C levels mg/kg				Linear effect	Quadratic effect
	60	250	500	750		
CR (%)[1] dp	88,78	90,39	90,59	87,59	ns	ns

	(± 4,03)	(± 0,94)	(± 0,46)	(± 5,11)		
RF (%)² dp	35,32 (± 0,94)	32,48 (± 1,15)	34,69 (± 2,94)	26,63 (± 6,15)	0,018	
IDS (%)³ dp	9,48 (± 0,64)	9,48 (± 0,64)	9,41 (± 0,46)	9,76 (± 0,55)	ns	ns
IHS (%)⁴ dp	2,24 (± 0,10)	2,07 (± 0,07)	1,97 (± 0,29)	2,02 (± 0,14)	ns	ns

(RC)¹: carcass yield; (RF)²: fillet yield; (IDS)⁽³⁾: somatic digestive index; (IHS)⁽⁴⁾: hepatosomatic index; sd (+ standard deviation);

Table 3 shows that the fillet yield averages resulted in a decreasing linear effect (y = 35.859 - 0.0095X; R^2 = 60.55%) as the level of vitamin C in the diet increased (Figure 4), with averages ranging from 35.32 to 26.63 %. The vitamin C levels of 250 and 500 mg / kg resulted in fillet yield averages similar to those obtained with the 60 mg / kg feed. However, for the 750 mg / kg level there was a decrease in fillet yield, which is probably due to the higher live body weight and consequent higher DMI, since the same trend was observed for carcass yield.

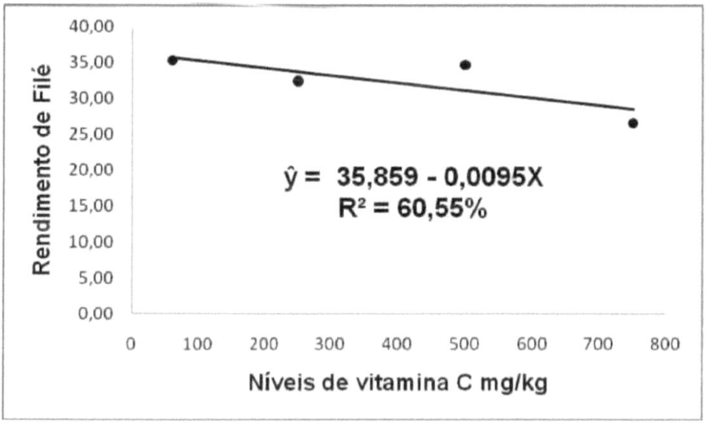

Figura 4. Fillet yield of très pintas piau fed supplementary levels of vitamin C.

According to the relevant literature, lighter fish tend to accumulate less visceral fat and provide a higher fillet yield, since the calculations were carried out according to the live body weight or whole fish, showing a direct interaction with the composition of the diet, growth physiology, standardization of the filleting technique, and others (CLEMENTS & LOVELL, 1994; MACEDO-VIÉGAS & SOUZA, 2004; LEONHARDT et al, 2006). However, the values obtained for fillet yield were close to those found with matrinx, which were 37.23 and 36.61% by GOMIERO et al., (2003), in agreement with NETO et al., (2006) and ALVES & POMPEU (2007).

In relation to the Somatic Digestive Index (SDI) and the Hepato-Somatic Index (HSI), no significant effect was found by the regression analysis (Table 3). Mathis et al. (2003) reported that high somatic digestive indices occur due to the consequent increase in viscera weight, which generally reduces carcass yield. In this study, the average values determined *ranged* from 9.41 to 9.76 %, and although no significant effect was found by the regression analysis, a certain tendency could be observed in *the L. renhardit* fish fed at the 500mg/kg level to provide lower SDI and higher CR. Variations found in carcass and fillet yield values are directly attributed to visceral weight and volume (LAZARRI et al., 2006). However, some caution should be exercised when interpreting the results, as nutritional levels and types of food are key factors in these changes, and not just the isolated effect of vitamin C supplementation, which suggests further research into the phenotypic or physiological expression of growth in three-spotted piau fish.

Table 4 shows the chemical composition of the fillets of très pintas piau fed supplementary levels of vitamin C, in which a significant linear effect was observed on the mineral matter content of the fillets ($\ddot{y} = 19.7370 + 0.0042X$; $R^2 = 98.70\%$) and a quadratic effect on the crude energy content of the fillets ($\ddot{y} = 5484.2080 + 0.3370X - 0.004X^2$; $R^2 = 96.15\%$).

Table 4: Averages of the chemical composition of the fillet of three-spotted piau fish fed supplementary levels of vitamin C (DM basis).

Fillet composition	Vitamin C levels mg/kg				Linear ■ effect	Quadratic effect
	60	250	500	750		
DM (%)[1] dp	96,69 (±0,54)	96,87 (± 0,38)	97,10 (± 0,58)	96,69 (± 0,30)	ns	ns
PB (%)[2] dp	81,00 (± 1,42)	80,04 (± 1,76)	80,58 (± 1,92)	80,80 (± 1,98)	ns	ns
EB (kcal/kg)[3] dp	5486,71 (± 0,09)	5416,08 (± 23,96)	5417,58 (± 44,88)	5441,16 (± 24,52)		0,028
MM (%)[4] dp	19,60 (± 1,30)	20,90 (± 1,14)	21,65 (± 1,55)	22,90 (± 0,49)	0,030	

(1) Dry matter; (2) Crude protein; (3) Crude energy; (4) Mineral matter; sd (± standard deviation);

The results presented in this study corroborate the PB values found by Garcia et al., (1999) which were also similar to those obtained by Gonçalves et al., (2002) with piavuçu (*L. macrocephalus*) fry and Iziel et al., (2004) with matrinxâ (*Brycon cephalus*) fry, when determining the protein composition of the fillet of these species. High levels of protein in the fillet indicate that there has probably been a reaction to the absorption of vitamin C in the animal organism, which implies an improvement in metabolic reactions, especially in the formation and regeneration of body tissue, with the hydroxylation of the amino acids proline and lysine, necessary for the formation of collagen present in connective tissues, also acting in the protein synthesis of collagen in the bone matrix (ROTTA, 2003).

The MM content of the chemical composition of the three-spotted piau fillet in this study can be considered high, probably due to the fact that the fish were not peeled moments before the skin was removed and filleted (contamination by scales in the fillet). However, the MM content obtained by Arbelàez-rojas et al. (2002) was comparable and close to that of the present study (Figure 5).

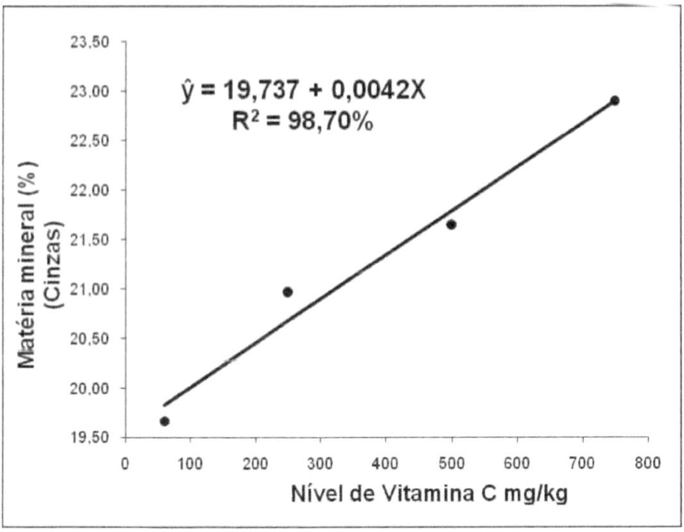

Figura 5. Percentage of ash in the fillet of three-spotted piau fed supplementary levels of vitamin C.

The differences observed in the crude energy values of the fillet raise questions about the use of vitamin C in fish feed, not only because of its effect on production performance but also because it improves the nutritional quality of the product. Although related studies are still in their infancy, it was observed in this work that the extra dietary levels added enabled a lower calorie deposition in the fillet *of L. reinhardit.*

These crudo energy values were similar to those found by Iziel et al. (2004), which ranged from 5750 to 6003 kcal / kg. These data are close to the results obtained in this study, which ranged from 5416 to 5486 kcal / kg.

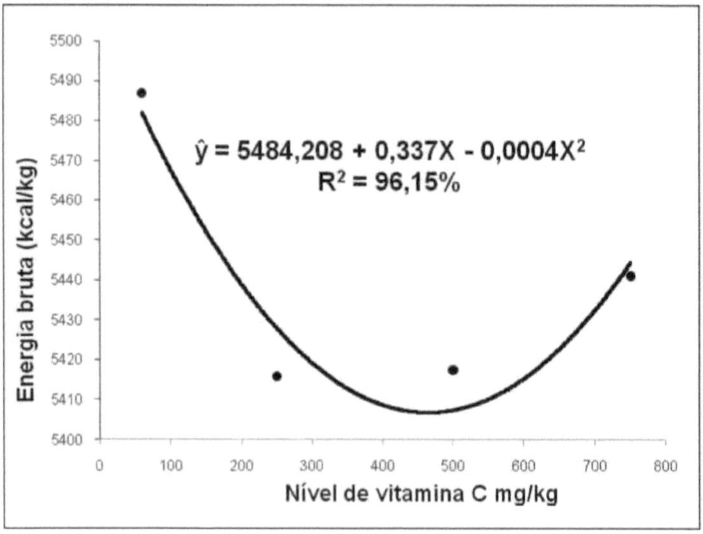

Figura 6. Crude energy of three-spotted piau fillets fed with supplementary levels of vitamin C.

Table 5 shows that the levels of vitamin C had a linear effect on the dry matter content of the carcass of juvenile three-spotted piau, showing a certain similarity in relation to the nutritional values obtained from the composition of the fillet, differing only in the dry matter content; however, this difference did not imply a significant variation in the other fractions of the carcass' chemical composition.

Table 5 - Averages of the chemical composition of the carcass of three-spotted piau fish fed supplementary levels of vitamin C in the feed (DM basis)

Carcass composition	Vitamin C levels mg/kg				Linear effect	Quadratic effect
	60	250	500	750		
DM (%)[1] dp	97,35 (± 0,88)	98,11 (± 0,07)	96,54 (± 0,27)	97,19 (± 0,22)	0,016	
PB (%)[2] dp	56,81 (±10,35)	54,03 (± 4,15)	52,03 (± 2,38)	57,50 (± 5,70)	ns	ns
EE (%)[3] dp	31,10 (± 7,21)	34,38 (± 0,87)	37,05 (± 3,94)	33,90 (± 1,03)	ns	ns
EB (kcal/kg)[4] dp	5786,07 (± 236,11)	5854,00 (± 126,51)	5871,54 (± 53,63)	5742,46 (± 53,94)	ns	ns
CZ (%)[5] dp	27,69 (± 9,88)	21,15 (± 2,52)	22,68 (± 1,84)	24,07 (± 1,30)	ns	ns

(1)Dry Matter; (2)Crude Protein; (3)Ethereal Extract; (4)Crude Energy; (5)Ash; sd(± standard deviation).

The DM content of the chemical composition of the carcass of three-spotted piau was the only variable that showed a significant difference ($p<0.05$) with a range of 96.54 and 98.11%. These values were similar to those found by SOUZA et al. (2004) who evaluated the carcass characteristics of piauçu fry and found a value of 89.48 % DM. On the other hand, Fujimoto et al., (2007) working with pacu juveniles reported values of 32% dry matter; Santos et al., (2006) studied the composition of palometa (*Serrassalmus spilopleura*) and found a value of 20.94% DM; and Gaiotto et al., (2004) with a similar value of 25.16% DM.

There was no significant effect on the deposition of ether extract and crude protein in the carcass of three-spotted piau. However, the lowest value of ether extract was found in the treatment of 60 mg of vitamin C / kg of feed and a value close to the highest content of crude protein found in the 750 mg level of vitamin C. It can be inferred that fish fed both 60 mg and extra dietary levels of vitamin C in their diets may favor greater protein deposition in this tissue. However, this hypothesis could only be confirmed with calculations of protein and energy retention, through body chemistry analysis of the fry at the beginning of the experiment. Martins (1994) showed that the highest level of vitamin C supplementation tested on *P. mesopotamicus*, under the conditions studied, improved final protein deposition in the carcass of these fish.

When evaluating the body composition of tambaqui, Arbelàez-rojas et al. (2002) found no significant difference in the fish's body PB and MM values, which was equivalent to the present study. Similar results were also obtained by (BITTENCOURT, 2008) working with pacu at different rearing densities. This suggests that knowledge of the metabolic responses of the *Leporinus* genus under different feeding conditions may make it possible to adjust the use of nutrients to improve growth and prepare more economical diets (Camilo et al., 2007).

Table 6 shows the results of the glycemic parameters in the blood and liver. There was no effect of treatment, but there was a significant effect in the analysis of variance ($p<0.05$) for blood glucose concentration, and liver glycogen showed the same pattern as blood glucose in relation to supplementary vitamin C levels.

Table 6. Glycemic parameters of très pintas piau fish fed supplementary levels of vitamin C in their diets.

Metabolic parameters	Vitamin C levels mg/kg				. Linear effect	Quadratic effect
	60	250	500	750		
GLI[1] dp	81,50 (± 14,9)	67,25 (± 2,2)	119,70 (± 44,6)	53,62 (± 12,0)	ns	ns
GCO[2] dp	1962,20 (± 1315,5)	1878,83 (± 166,4)	2128,25 (± 1592,7)	1264,38 (± 865,6)	ns	ns

(GLI)(1) = Blood glucose (mg/dL); (GCO)(2) = Hepatic glycogen (μmol glycosyl-glucose/g).

The blood samples taken to measure the glucose concentration were taken at the time of harvest, which preceded slaughter, so it can be inferred that the lower average value of 53.62 mg / dL for the glucose level of the fish fed the supplementary vitamin C level of 750 mg / kg was effective in containing the glucose rise caused by the stress of capture at the time of harvest. However, it is interesting to note that the rise in glucose levels was 119.70 mg / dL in the fish fed 500 mg / kg. This may have occurred because the fish were agitated at the time of harvest and spent more time in the bucket. Similar values were found by Chagas et al., (2007), who found blood glucose levels of 60.09 mg / dL in tambaqui fry, while Okamura et al., (2007) found lower blood glucose values with the 500 and 1000 mg levels of vitamin C fed to nilotic tilapia. (HENRIQUE et al., 1998 and ORTUNO et al., 2003) also found significant efficiency in containing glucose rises in fish supplemented with ascorbate under stress conditions.

The increase in blood glucose and glycogen in the liver characterizes biochemical changes in the glucose and gluconeogenesis pathways to maintain the processes of obtaining energy, probably due to the difficulty at the time of capture (agitation and movement), causing these animals stress until the time of slaughter.

The stress suffered by these animals during harvest after 24 hours of fasting was representative in the fish treated with 500 mg of vitamin C/kg of feed. In this case, the amount of vitamin C available in the tissues was not enough to minimize this response. It can also be inferred that the action of the ascorbate reductase enzyme in preventing the loss of this vitamin in the cells was possibly inefficient, due to the lower possibility of dehydroascorbic acid hydrolysis in the organism of the three-spotted piau. As a physiological response, the body tends to increase glucose levels to meet the energy demand. This effect may also have occurred due to alterations in cortisol biosynthesis, as vitamin C is involved in the prevention and conversion of unsaturated fatty acids into cholesterol esters, which are important components of this hormone (MONTERO et al., 1999), as Carneiro et al. (2002) point out that cortisol can help maintain high blood glucose levels after stress.

Façanha & Gomes (2005) working with juvenile tambaqui exposed to different concentrations of menthol found measurements ranging from 63.20 to 100.50 mg dL of glucose, showing that the higher the dosage, the higher the blood glucose concentration. In the same year, Inoue et al. (2005) found glucose values ranging from 60 to 120 mg dL and, consequently, hepatic glycogen levels ranging from 350 to 550 µmol / g were observed in relation to the transportation of live fish, which is certainly one of the main adverse stimuli to fish homeostasis in captive breeding conditions.

Brandâo et al., (2006) studied the stress response in pirarucu (*Arapaima gigas*) during routine fish farming practices and observed that glucose varied from 20 to 99.5 mg dL. In a trial with matrinx fed different levels of vitamin C and subjected to exposure to air (ABREU

& URBINATI, 2006) they found blood glucose levels ranging from 90 to 120 mg dL, while liver glycogen levels showed no difference to the levels of vitamin C tested.

In this study, the performance variables and physiological parameters did not show significant effects or differences, but certain benefits can be inferred for extra dietary vitamin C supplementation in the diets, highlighting that the complexity of the zootechnical exploitation environment, nutritional and environmental interactions should be further clarified to better understand the requirements and recommendations of vitamin C for three-spotted piau in intensive systems.

2. 4 CONCLUSION

The level of vitamin C added to commercial feed is adequate for three-spotted piau fry. The extra dietary supplementation of vitamin C in the diet during the growth phase (fry-juvenile) of the three-spotted piau showed benefits for zootechnical indices and the nutritional composition of the fillet and carcass, as well as reducing the effect of production and/or harvest stress.

REFERENCES

ABREU, J. S.; URBINATI, E. C. Physiological responses of matrinxa (*Brycon amazonicus*) fed different levels of vitamin C and submitted to air exposure. **Acta amazônica**. v. 36, n. 4, p. 519-524, Oct. 2006.

AI, Q. ET AL. Effects of dietary vitamin C on growth of immune response of Japanese seabass (*Lateolabrax japanicus*) **Aquiculture**, v. 242, p. 489-500, 2004.

ANDERSEN, F. ET AL. Interaction between two dietary levels of iron and two forms of the ascorbic acid and the effect on growth, antioxidant status and some non-specific immune parameters in atlantic salmon (*Salmo salar*) smolts. **Aquaculture,** Amsterdam, v. 161, p. 437-451, 1998.

ARBELAEZ-ROJAS, G. A.; FRACALOSSI, D. M.; FIM, J. D. I. Composición corporal de tambaqui, (*Colossoma macropomum*) e matrinxa (*Bricon cephalus*), em sistemas de cultivo intensivo, em igarapé, e semi-intensivo, em viveiros. **Revista Brasileira de Zootecnia**. v. 31, n. 3, p. 17. Viçosa, Jun. 2002.

ALMEIDA, G. S. C. **Dietary supplementation of vitamin C, development and health of pacu (*Piaractus mesopotamicus*- Holmborg, 1887)**. p. 60. 2003. Dissertation (Master's Degree) - Piracicaba, SP. 2003.

ALVES, M. F.; LEÂO, F. P.; POMPEU, P. S. Feeding of the three-spotted piau (*Leporinus *reinhardit*) - Lutken, 1874, (Anostomidae, Characiformes). 2007. **ANIAS...** VIII Congress of Ecology of Brazil, C Axambu-MG. 2007.

ASSOCIATION OF OFFICIAL AND AGRICULTURAL CHEMISTRY. **Official Methods of Analysis**. 5th. Washington, D.C. USA, 1990.

BACCONI, D. F. **Nutritional requirement of vitamin A for Nile tilapia (*Oreochromis niloticus*) fry.** Dissertation (Master's Degree). Luiz de Queiroz College - ESALQ, Piracicaba, SP. 2003.

BIDINOTTO, P. M.; SOUZA, R. H. S.; MORAES, GHepatic glycogen in eight tropical freswater teleost fish: Aprocedure for Field determinations of microsamples. **Technical Bulletin-CEPTA,** 1998, n. 10, p. 53-60.

BITTENCOURT, F. **Cultivation of pacu (*Piaractus mesopotamicus*) at different densities in tank-nets in the Itaipu reservoir**. p. 46. 2008. Dissertation (master's degree), Marechal Cândido Rondon, 2008.

BARROS, M. M.; PEZZATO, L. E.; KLEEMANN, G. K.; HISANO, H.; ROSA, G. J. M. Levels of vitamin C and iron for Nile tilapia (*Oreochromis niloticus)* **Revista Brasileira de Zootecnia**, v.31, n. 6, p. 2149-2156, Jul. 2002.

BRANDÂO, F. R.; GOMES, L. C.; CHAGAS, E. C. Stress responses in pirarucu (*Arapaima gigas*) during routine fish farming practices. **Acta Amazônica,** v. 36, n.3, p. 349-356, Aug. 2006.

CAMILO, R. Y.; HONORATO, C. A.; MORAES, G. Effect of fasting, absence of oil in the diet and dietary supplementation with essential amino acids on the intermediary metabolism of *Brycon amazonicus*. In: 1st BRAZILIAN CONGRESS ON NATIVE SWEETWATER FISH PRODUCTION. 1ST MEETING OF FISH FARMERS FROM MATO GROSSO DO SUL. **ANNALS...** Mato Grosso do Sul. 2007.

CARNEIRO, P. C. F.; MIKOS, J. D.; BENDHACK, F.; IGNASIO, S. A. Processing of jundià (*Rhamdia quelen)* carcass yield. **Revista acadêmica**: Ciências Agràrias e Ambientais, Curitiba, v.2, n. 3, p. 11-17. March. 2004.

CAVICHIOLO, F.; VARGAS, L.; RIBEIRO, R. P.; MOREIRA, H. L.; LOURDES, B. R. R.; MAEHANA, K. P. J. A.; LEONARDO, J. M. L. O. Effect of vitamin C and E supplementation in the diet on the occurrence of ectoparasites, performance and survival in Nile Tilapia (*Oreochromis niloticus, L.*) larvae during sex reversal. **Acta scientiarum.** v. 24, n. 4, p. 943 - 948, Maringà. July. 2002a.

COSTA, L. D. F.; MIRANDA-FILHO, K. C.; SEVERO, M. P.; SAMPAIO, L. A. Tolerance of juvenile pompano *Trachinotus marginatus* to acute ammonia and nitrite exposure at different salinity levels. **Aquaculture**, Amsterdan, 2008., v. 285. n. 4. p. 270-272. 2008.

CHAGAS, E. C.; VAL, A. L. Effect of vitamin C on weight gain and hematological parameters of tambaqui. **Pesquisa Agropecuària Brasileira,** v. 38, n. 3, Brasilia. 2003.

CHAGAS, E. C.; GOMES, L. C.; JUNIOR, H. M.; ROUBACH, R. Productivity of tambaqui reared in net pens with different feeding rates. **Ciência Rural**, Santa Maria, v. 37, n. 4, p. 1109-1115. 2007.

CLEMENT, S.; LOVELL, R. T. Comparison of processing yeld and nutrient composition of cultured nile tilapia (*Oreochromis niloticus*) and channel catfish (*Ictalurus punctatus*). **Aquaculture,** v. 119, p. 299-310. 1994.

DUBOIS, M. G.; GILLES, K. A.; HAMILTON,J. K.; REBERS, P. A.; SMITH, F. Colorimetric method for determination of sugars and related substances. **Analisys chemisttry,** n.28, p. 350-358. 1960.

FAÇANHA, M. F.; GOMES, L. C. The effectiveness of menthol as an anesthetic for tambaqui (*Colossoma macropomum*) CHARACIFORMES, CHARACIDAE. **Acta Amazônica**. v. 35, n. 1, p. 71-75, 2005.

FRACALOSSI, D. M.; ALLEN, M. E.; NICHOLS, D. K.; OFTEDAL, O. T. Oscars (*astronotus ocellatus*), have a dietary requirement for vitamin c. **Journal of nutrition,** Bethesda, v. 128, p. 1745 - 1751. 1998.

FUJINOTO, R. Y.; CASTRO, M. P.; HONORATO, C. A.; MORAES, F. R. Body composition and nutrient utilization efficiency for pacus fed a diet supplemented with trivalent chromium. **Revista pesquisa agropecuària brasileira**, Brasilia, v. 42, n. 12, p. 1763-1768, Dec. 2007.

FUJIMOTO, R. Y., CARNEIRO, D. J. Addition of ascorbyl polyphosphate as a source of vitamin C in diets for pintado fry, (*Pseudoplatystoma corruscans)* Agassiz - 1829. **Acta Sci. Animal Sci,** Maringà, v. 23, n. 4, p. 855-861.2001.

GAIOTTO, J. R.; MACEDO-VIEGAS, E. M.; FERNANDES, T. R. Canola meal for juvenile Nile tilapia (*Oreochromis niloticus*), chitralada strain. **Acta scientiarum. Animal sciences**, Maringá, v. 26, n. 1, p. 15-19, 2004.

GARCIA, J. E.; PEZZATO, L. E.; FILHO, E. Z.; VICENTINI, C. A. Utilization of crude fiber in the nutrition of piracanjuba (*Brycon orbignyanus*). **Acta Scientiarum**, v. 21, n. 3, p. 725-731. 1999.

GOMIERO, J. S. G.; RIBEIRO, P. A. P.; FERREIRA, M. W.; LOGATO, P. V. R. Carcass yield of matrinxa fish (*Bricon cephalus*) in the different head cuts. **Ciência Agrotecnologica**, Lavras, v. 27, n. 1, p. 211-216. Jan/Feb. 2003.

GONÇALVES, G. S.; FURUYA, W. M.; RIBEIRO, R. P.; FURUYA, V. R. B.; SOARES, C. M. Canola bran in the diet of the early stage fish, *Leporinus macrocephalus* (Garavello & Britski). **Acta scientiarum**, Maringà, v.24, n.4, p. 921-925. 2002.

GURGEL, J. J. S.; FREITAS, J. V. F. Sobre a composiçao quimica de doze espécies de peixes de valor comercial de açudes do nordeste brasileiro. **Boletim técnico**, Fortaleza, v. 30, n.1, p. 45-57, jav/ jun, 1972.

HENRIQUE, M. M. F.; GOMES, A. E. F.; GOUILLOU-COUSTANS, M. F. et al. Influence of supplementation of practical diets with vitamin C on growth and response to hypoxic stress of seabream (*Sparus aurata*). **Aquaculture,** v.161, p. 415-426. 1998.

INOUE, L. A. K. A.; AFONSO, L. O. B.; IWAMA, G. K.; MORAES, G. Effects of clove oil on the stress response of matrinxa (*Bricon cephalus*) subjected to transport. **Acta Amazônica**, v. 35, n. 2, p. 289-295, Maio. 2005.

IZIEL, A. C. U.; FILHO-PEREIRA, M.; MELLO, S. A. S.; MACEDO, J. L.V. Evaluation of protein levels for the nutrition of juvenile matrinx (*Brycon cephalus*). **Acta amazônica,** v. 34, n. 2, p. 179-184. 2004.

JOBLING, M.; KOSKELA, J.; SAVOLAINEN, R. Influence of dietary fat level and increased diposity on growth and fat deposition in rainbow trout, *Oncorhynchus mykiss* (Walbaum). **Aquaculture Research**, v. 29, p.601-607. 1998.

KASAI, R. Y. D. **Vitamin C in diets for trairao fry**. p. 55, 2007. Dissertation (master's degree), Federal University of Viçosa, 2007.

KUBOTA, E.H; EMANUELLI, T. Fish processing. In: BALDISSEROTTO, B. NETO, J, R. **Criaçao de Jundiâ**. Santa Maria: UFSM, 2004, p.117-141.

LAZARRI, R. et al. Different protein sources for feeding jundiâ (*Rhamdia quelen*). **Ciência Rural**, v. 36, n. 1, p. 240-246. 2006.

LEHNINGER, A. L.; NELSON, D. L.; COX, M. M. **Principios da bioquimica**. Sao Paulo: Sarvier, 1995.

LEONHARDT, J. H.; FILHO, M. C.; FROSSARD, H.; MORENO, A. M. Morphometric characteristics, yield and composition of the fillet of Nile tilapia (*Oreochromis niloticus*), local Thai strain and crossbreed of both. **Semina: Ciências Agràrias**, Londrina, v. 27, n. 1, p.

125-132. 2006.

LUNA- FIGUEROA, J. Influencia de alimento vivo en La reproducción y crescimento Del pez Angel (*Pterophyllun scalare*) - Pisces: Cichlidae, **Revista Acta Unicersitaria**, v. 9, n. 2, p. 34-40. 1999.

MACEDO-VIEGAS, E. M.; SOUZA, M. L. R. Pre-processing and conservation of fish produced in fish farms. IN: CYRINO, J. E. P. ET AL., (ED) Tópicos espéciais em piscicultura de aqua doce tropical intensiva. SÂO PAULO: Tecart, 2004, p. 405-500.

MARTINS, M. L. **Effects of ascorbic acid (vitamin C) supplementation evaluated through performance and anatomopathological aspects in fry of *Piaractus mesopotamicus* - Holmberg, 1887**. Dissertation (Master's Degree) - Jaboticabal, SP. 1994.

MATHIS, N.; FEIDT, C.; BRUN-BELLUT, J. Influence of protein / energy ratio on carcass quality during the growing period of Eurasian porch (*Perça fluviatilis*). **Aquaculture,** v. 217, p. 453-464. 2003.

MELLO, R. F.; MOURA, M. A. M.; VIEIRA, I.; CYRINO, J. E. P. Supplementing the diet of piauçu (*Leporinus obtusidens*) fry with vitamin C. **Scientia Agricola**, v. 56, n. 4, p. 1223-1231, Suplemento. 1999.

MIRANDA, E. C.; PINTO, L. G. Q.; FURUYA, W. M.; PEZZATO, L. E.; BARROS, M. M.; PEZZATO, A. C. Weight gain and survival rate of pacu (*Piaractus mesopotamicus*) post-larvae fed diets containing different levels of vitamin C. **Acta Scientiarum Animal Sciences**, Maringà, v. 25, n. 1, p. 31-36. 2003

MONTERO, D.; MARRERO, M.; IZQUIERDO, M. S.; ROBAINA, L.; VERGARA, J. M.; TORT, L. Effect of vitamin E and C dietary supplementation on some immune parameters of gilthead seabream (*Sparus aurata*) juveniles subjected to crowding stress. **Aquaculture**, v. 171, p. 269-278, 1999.

MUZITANO, I. S. **Effect of vitamin C supplementation on the performance of melanotenia maça (*Glossolepis incisus*).** Dissertation (Master's Degree) - Campos dos Goytacazes, RJ. 2007.

NATIONAL RESEARCH COUNCIL (NRC). Nutrient requirements of wamwater fishes and shellfishes. Washington: National Academy Press, 1993, p. 102.

NAVARRO, R. D.; LANNA, E. A. T.; DONZELE, J. L.; MATTA, S. L. P.; SOUZA, M. A. Dietary digestive energy levels on the performance of post-larval piauçu (*Leporinus *macrocephalus*). **Acta Sci. Anim. Sci.** Maringà, v. 29, n. 1, p. 109114. 2007.

NETO, J. R.; LAZARRI, R.; PEDRON, F. A.; VEIVERBERG, C. A.; BERGAMIN, G. T.; CORRÊIA, V.; FILIPETTO, J. E. S. Feeding of piava (*Leporinus obtusidens)* with different protein sources. **Ciência Rural**, Santa Maria, v. 36, n. 5, p. 1611-1616. Oct. 2006.

OKAMURA, D.; ARAUJO, F.G.; LOGATO, P.V.R.; MURGAS, L.D.S.; FREITAS, R.T.F.; ARAUJO, R.V. Effects of vitamin C on hematocrit and glycemia of Nile tilapia (*Oreochromis niloticus*) fry in simulated transportation. **Arch. Bràs. Med. Vet. Zootec.,** v. 59, n.4, p. 883-888. 2007.

OKAMURA, D.; ARAUJO, F. G.; LOGATO, P. V. R.; SILVEIRA, U. S.; MURGAS, L. D. S.; FREITAS, R. T. F. Ascorbyl palmitate and tocopherol acetate as metabolic antioxidants in gilthead seabream larvae. **Pesquisa Agropecuària Brasileira,** Brasilia, v. 43, n. 8, p. 1061 - 1068. Jun. 2008.

OMENA, C. M. B. **Reflexos da utilização de farelo** de **coco na alimentação de Tilàpia-do-Nilo (*Oreochromis niloticus* - LINNAUES, 1857) sobre *o* valor nutricional do filé**. p. 95, 2008, Dissertaçâo (Mestrado), Universidade Federal de Alagoas, 2008.

OGAWA, M.; MAIA, E. L. **Manual de pesca**: ciência e tecnologia do pescado. Sao Paulo: Varela, 1999. v. 1, p. 430.

ORTUNO, J.; ESTEBAN, J.; MESEGUER, M. A. The effect of dietary intake of vitamin C and E on the stress response of gilthead seabream (*Sparus aurata L.*) **Fish Shellfish Immunol,** v.14, p. 145-156. 2003.

PIZANGO-PAIMA, E. G. **Estudo da alimentação e composição corporal do matrinxâ, *Brycon cephalus* (Gunther, 1869) (Characiformes, Characidae) na Amazônia central**. Manaus: INPA/ FUA, p. 71. 1997. Dissertation (Master's Degree), 1997.

ROTTA, M. A. **Utilization of ascorbic acid (vitamin C) by fish**. Corumbà: Embrapa pantanal, 2003. p. 54, (Documentos / Embrapa pantanal, ISSN 1517-1973; 49), 2003.

SAS Institute (Cary, United States) Statistical Analysis System: User's Guide. United States: SAS INSTITUTE, 2002, p. 846.

SOLIMAN, A. K.; JAUNCEY, K.; ROBERTS, R. J. The effect of varying forms of dietary ascorbic acid on the nutrition of juvenile tilapia (*Oreochromis niloticus)*. **Aquaculture**, v. 52, p. 1-10. 1986.

SANTOS, A. B.; GINAR, R. M. B.; GALARÇA, R. C. G.; NETO, J. B. Composiçâo bromatologica do filé da palometa (*Serrasalmus spilopleura* Kner, 1860) na regiâo de

Uruguaiana - RS, Brasil, **Revista da FZVA**, Uruguaiana, v. 13, n. 2, p. 166-170. 2006.

SILVA, S. S.; ANDERSON, T. A. **Fish nutrition in aquiculture**. London: Chapman & Hall, p. 319, 1995.

SILVA, D. J.; QUEIROZ, A. C. Anàlise de alimentos métodos quimicos e biológicos. 3 ed. Viçosa, MG: Editora UFV, p. 235, 2002.

SOUZA, S. R.; HAYASHI, C.; SOARES, T.; ANDRADE, L. S. Evaluation of the effect of different levels of cottonseed meal on the performance and body composition of piavuçu fry (*Leporinus macrocephalus*). **Instituto brasileira de pesca,** Sao Paulo, v. 30, n. 2, p. 127-134, 2004.

SOARES, C. M.; HAYASHI, C.; FURUYA, V. R. B.; FURUYA, W. M.; GALDIOLLI, E. M. Partial and total substitution of soybean meal protein by canola meal protein in the feeding of piavuçu fry (*Leporinus macrocephalus, L.*) **Revista Bras. Zootec**. v. 29, n. 1, p. 15-22. 2000.

TACON, A. G. L. Vitamin nutricion in shrimp and fish. In: Aquacultures feed processing and nutrition, **Workshop**, Singapore.1991.

TOYAMA, G. N.; CORRENTE, J. E.; CYRINO, J. E. P. Vitamin C supplementation in Nile tilapia sex reversal actions. **Scientia Agricola,** v. 57, n. 2, p. 221-228. 2000.

THOMPSON, I. et al. The effect of stress on the immune response of atlantic salmon (*Salmo salar L.)* fed diets containing different amounts of vitamin C. **Aquaculture**, Amsterdam, v. 114, p. 1-18, 1993.

ZUANON, J. A. S.; SALARO, A. L.; BALBINO, E. M.; SARAIVA, A.; QUADROS, M.; FONTANARI, R. L. Levels of crude protein in diets for flagwhelk fry. **Revista Brasileira de Zootecnia,** v. 35, n.5, p. 1893-1896. 2006.

FINAL CONSIDERATIONS

Brazil has the richest diversity of fish species with great economic potential. *L. reinhardit* was chosen as an experimental animal because it is native to the São Francisco River and there is little information about it. However, this little information reports its great potential for zootechnical performance.

Another fact that must be taken into account is the use of nutrients in feed, such as vitamins, lipids and so on. Vitamin C is a nutrient that is often used as a stimulant and performance promoter. Animals fed diets deficient in vitamin C show physical deformities, slow healing of injuries and increased mortality. Nowadays, vitamin C is commonly used, but the lack of a table of Brazilian nutritional requirements for the industry and fish farmers encourages the use of high and variable amounts of this vitamin in feed, making the cost-benefit ratio for production systems questionable.

According to the results obtained, it can be concluded that the extra dietary supplementation of vitamin C in the growth diet of the très pintas piau brings benefits for the rearing of this fish in intensive production systems. However, the scarcity of information makes it difficult to compare the results regarding the action of vitamin C in the organism of these species.

It is therefore essential that the results of research carried out in this area, as in any other, are well publicized so that information can be exchanged and complemented. Only in this way will it be possible to establish the appropriate level for each native species, which will facilitate further studies and promote the advancement of research into these species.

ANNEXES

Figura 7. Field experiment site, COD EVASF headquarters, experimont installation shed, 500L experimental aquariums, fry reception tanks.

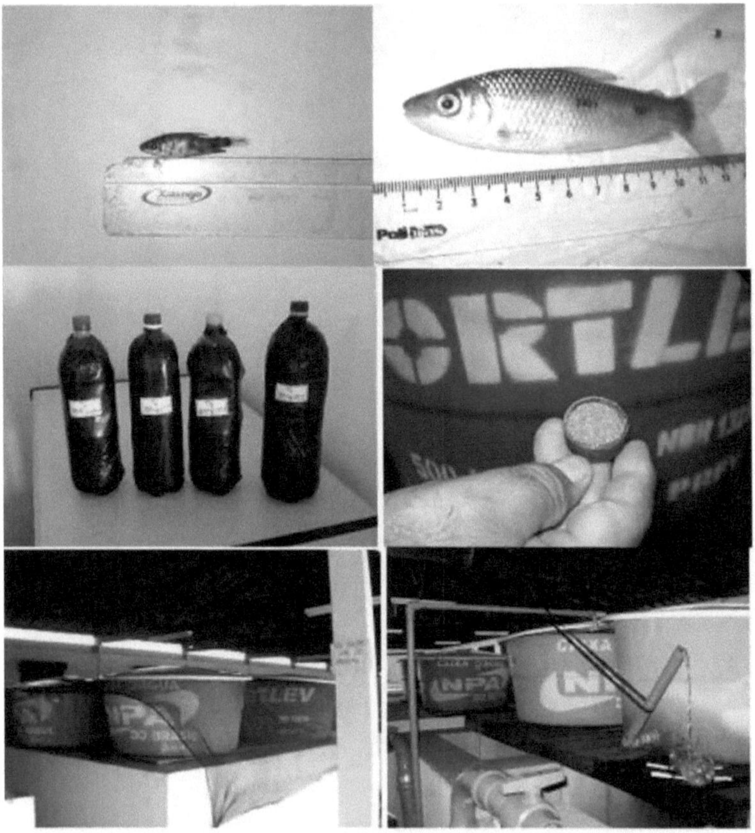

Figura 8. Initial fry size, final fry size, experimental feed storage containers, feed distribution containers, experimental aquariums and constant water circulation system.

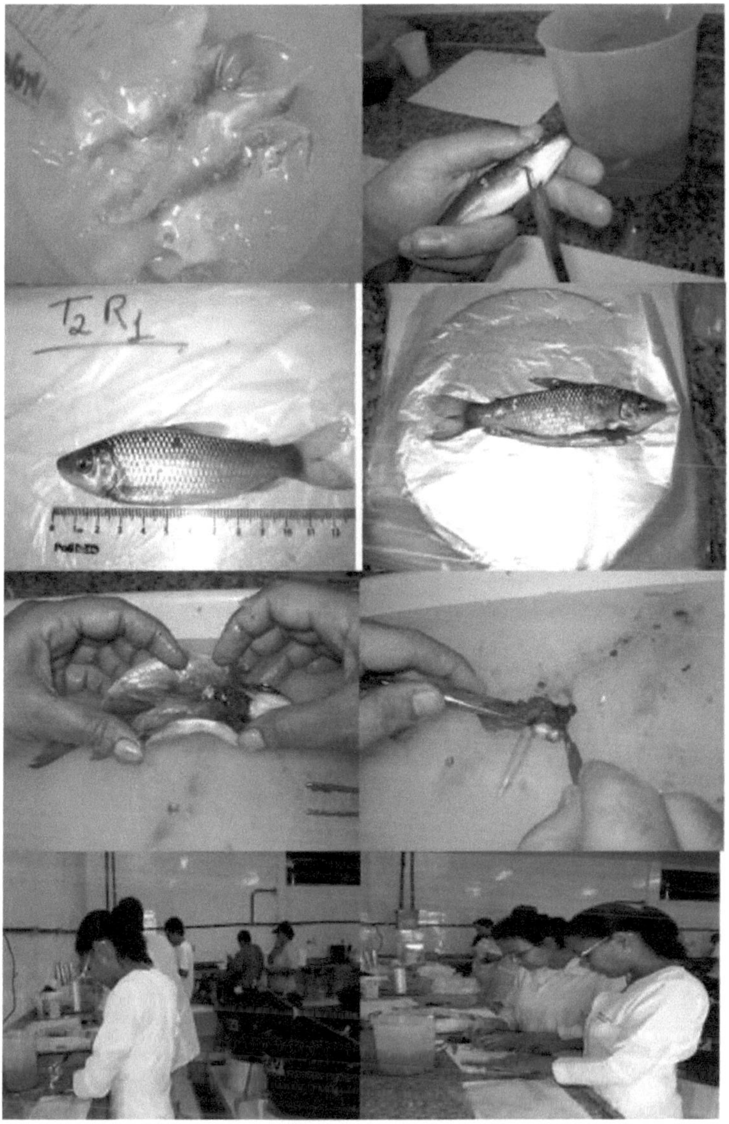

Figura 9. Slaughtering the animals, removing the blood, measuring the length, weighing, removing the viscera, separating the liver and the LENAB work team.

MIX
Papier aus verantwortungsvollen Quellen
Paper from responsible sources
FSC® C105338

Printed by Books on Demand GmbH, Norderstedt / Germany